Student Solutions Manual for

Finite Mathematics

Practical Applications

Michael Rosenborg

BROOKS/COLE PUBLISHING COMPANY

I(T)P® An International Thomson Publishing Company

Pacific Grove • Albany • Belmont • Bonn • Boston • Cincinnati • Johannesburg • London
Madrid • Melbourne • Mexico City • New York • Scottsdale • Singapore • Tokyo • Toronto

Assistant Editor: *Melissa Duge*
Editorial Assistant: *Shelley Gesicki*
Marketing Representative: *Caroline Croley*
Marketing Assistant: *Debra Johnston*
Production Editor: *Mary Vezilich*
Cover Design: *Roy Neuhaus*
Printing and Binding: *Webcom Limited*

For more information, contact:

BROOKS/COLE PUBLISHING COMPANY
511 Forest Lodge Road
Pacific Grove, CA 93950
USA

International Thomson Publishing Europe
Berkshire House 168-173
High Holborn
London WC1V 7AA
England

Thomas Nelson Australia
102 Dodds Street
South Melbourne, 3205
Victoria, Australia

Nelson Canada
1120 Birchmount Road
Scarborough, Ontario
Canada M1K 5G4

International Thomson Editores
Seneca 53
Col. Polanco
11560 México, D. F., México

International Thomson Publishing Japan
Hirakawacho Kyowa Building, 3F
2-2-1 Hirakawacho
Chiyoda-ku, Tokyo 102
Japan

International Thomson Publishing Asia
60 Albert Street
#15-01 Albert Complex
Singapore 189969

International Thomson Publishing GmbH
Königswinterer Strasse 418
53227 Bonn
Germany

Printed in Canada

5 4 3 2 1

ISBN 0-534-93408-0

Contents

Chapter 1
Linear Equations

Section 1.0
Lines and their Equations

1a. Rise = $y_2 - y_1 = 11 - 7 = 4$; run $= x_2 - x_1 = 5 - 3 = 2$; slope $= \frac{\text{rise}}{\text{run}} = \frac{y_2 - y_1}{x_2 - x_1} = \frac{4}{2} = 2$.

1b. Rise $= y_2 - y_1 = 7 - 11 = -4$; run $= x_2 - x_1 = 3 - 5 = -2$; slope $= \frac{\text{rise}}{\text{run}} = \frac{y_2 - y_1}{x_2 - x_1} = \frac{-4}{-2} = 2$. *Note:* $\frac{-4}{-2} = \frac{-1(4)}{-1(2)} = \frac{-1}{-1} \cdot \frac{4}{2} = 1 \cdot 2 = 2$.

1c. We can conclude that the order in which the points are considered does not matter when computing slopes. That is, if we switch the points given for (x_1, y_1) and (x_2, y_2), we will arrive at the same slope. This was actually done in **1a** and **1b** above.

1d.

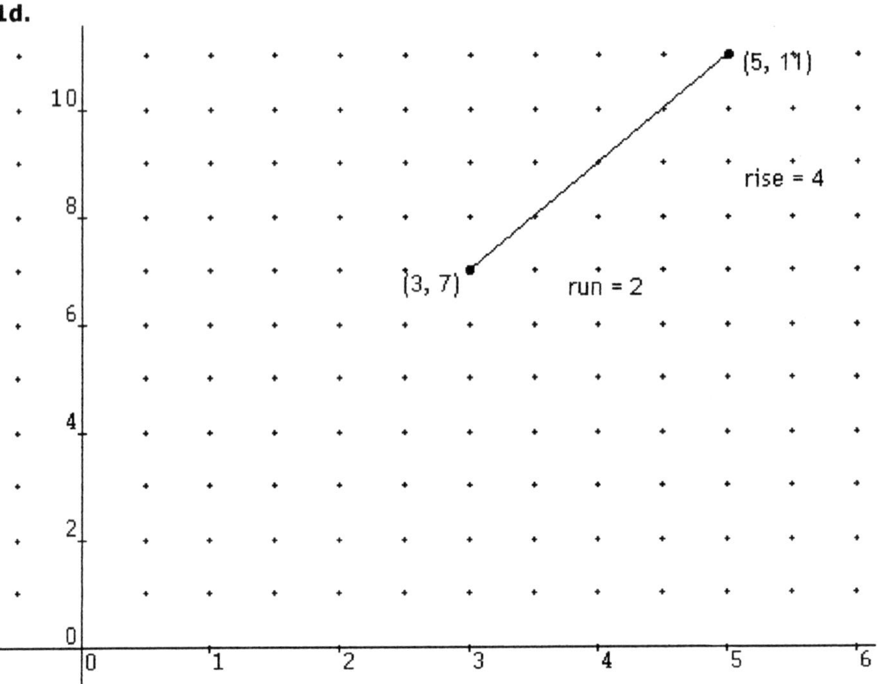

5a.

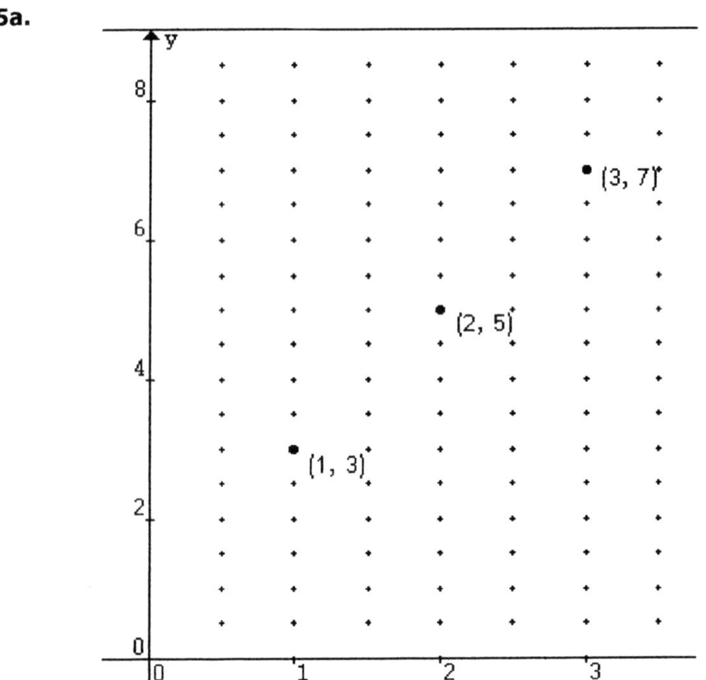

5b. Let $(x_1, y_1) = (1, 3)$, $(x_2, y_2) = (2, 5)$. Then

$$\text{slope} = \frac{y_2 - y_1}{x_2 - x_1} = \frac{5-3}{2-1} = \frac{2}{1} = 2.$$

5c. Let $(x_1, y_1) = (2, 5)$, $(x_2, y_2) = (3, 7)$. Then

$$\text{slope} = \frac{y_2 - y_1}{x_2 - x_1} = \frac{7-5}{3-2} = \frac{2}{1} = 2.$$

5d. Yes, the three points are in a line because the first two and second two points of the three points have the same slope.

9a.

0
−8 −6 −4 −2 0 2 4
−5
−10
(4, -2)
(-4, -8)
(-8, -11)

9b. Let $(x_1, y_1) = (-8, -11)$, $(x_2, y_2) = (-4, -8)$. Then

$$\text{slope} = \frac{y_2 - y_1}{x_2 - x_1} = \frac{-8-(-11)}{-4-(-8)} = \frac{-8+11}{-4+8} = \frac{3}{4}.$$

9c. Let $(x_1, y_1) = (-4, -8)$, $(x_2, y_2) = (4, -2)$. Then

$$\text{slope} = \frac{y_2 - y_1}{x_2 - x_1} = \frac{-2-(-8)}{4-(-4)} = \frac{-2+8}{4+4} = \frac{6}{8} = \frac{3}{4}.$$

9d. Yes, the three points are in a line because the first two and second two points of the three points have the same slope.

13a. Let $(x_1, y_1) = (-6, -3)$, $(x_2, y_2) = (2, 4)$. Then

$$\text{slope} = \frac{y_2 - y_1}{x_2 - x_1} = \frac{4-(-3)}{2-(-6)} = \frac{4+3}{2+6} = \frac{7}{8}.$$

13b. Using the point $(2, 4)$ and the slope $\frac{7}{8}$, we have a rise of 7, which means we add 7 to the y-coordinate of the point:

$$7 + 4 = 11.$$

This is the y-coordinate of our new point. We have a run of 8, which means we add 8 to the x-coordinate of the point:

$$8 + 2 = 10.$$

This is the x-coordinate of our new point. That is, our new point on the same line is $(10, 11)$.

13c.

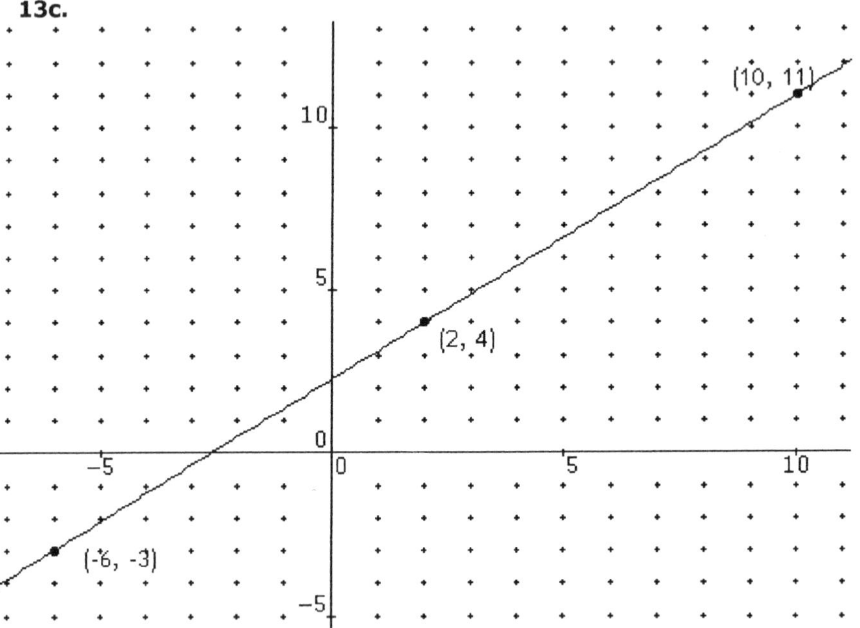

17. By using the slope between two points to make predictions, we implicitly assume that the relationship between price and cones sold is linear, but this is probably unjustified because as we all know, people and their buying behavior are unpredictable and hence may not follow a linear pattern at all. (Such assumptions are often necessary in textbooks for instructional purposes.) However, predictions following linear or other simple mathematical models may be reasonably made if much data over a long period of time is used; this is properly studied in statistics.

21a. We can combine the formula for the slope of a line between two points and the point-slope formula for a line to obtain a convenient formula for computing the equation of a line through two points as follows.

Recall that slope $m = \frac{y_2 - y_1}{x_2 - x_1}$, and the point-slope formula for a line is $y - y_1 = m(x - x_1)$, where (x_1, y_1), (x_2, y_2) are two given points on the line. Simply substitute the formula for slope m into the point-slope formula:

$$\begin{aligned} y - y_1 &= \boldsymbol{m}(x - x_1) \\ y - y_1 &= \frac{\boldsymbol{y_2 - y_1}}{\boldsymbol{x_2 - x_1}}(x - x_1) \end{aligned}$$

(Yes, the m in the point-slope formula represents slope, and is the same m in the slope formula.)

We use the formula we developed to find the equation of the line through the points $(x_1, y_1) = (-6, -3)$, $(x_2, y_2) = (2, 4)$:

$$\begin{aligned} y - y_1 &= \frac{y_2 - y_1}{x_2 - x_1}(x - x_1) \\ y - (-3) &= \frac{4 - (-3)}{2 - (-6)}[x - (-6)] \\ y + 3 &= \frac{4 + 3}{2 + 6}(x + 6) \\ y + 3 &= \frac{7}{8}(x + 6) \\ y + 3 &= \frac{7}{8}x + \frac{7}{8} \cdot 6 \\ y + 3 &= \frac{7}{8}x + \frac{21}{4} \\ y + 3 - 3 &= \frac{7}{8}x + \frac{21}{4} - 3 \\ y &= \frac{7}{8}x + \frac{21}{4} - 3 \cdot \frac{4}{4} \\ y &= \frac{7}{8}x + \frac{21}{4} - \frac{12}{4} \\ y &= \frac{7}{8}x + \frac{9}{4}, \end{aligned}$$

as desired.

21b. Let $(x, y) = (-6, -3)$. We substitute this point into $y = \frac{7}{8}x + \frac{9}{4}$:

$$\begin{aligned} -3 &= \frac{7}{8}(-6) + \frac{9}{4} \\ -3 &= -\frac{21}{4} + \frac{9}{4} \\ -3 &= -\frac{12}{4} \\ -3 &= -3 \end{aligned}$$

The point checks. Now let $(x, y) = (2, 4)$:

$$\begin{aligned} 4 &= \frac{7}{8}\cdot 2 + \frac{9}{4} \\ 4 &= \frac{7}{4} + \frac{9}{4} \\ 4 &= \frac{16}{4} \\ 4 &= 4 \end{aligned}$$

This point also checks.

21c. Let $(x, y) = (10, 11)$. We substitute this point into $y = \frac{7}{8}x + \frac{9}{4}$:

$$\begin{aligned} 11 &= \frac{7}{8}\cdot 10 + \frac{9}{4} \\ 11 &= \frac{35}{4} + \frac{9}{4} \\ 11 &= \frac{44}{4} \\ 11 &= 11 \end{aligned}$$

The point checks.

25a. Let $(x_1, y_1) = (8, 11)$, $(x_2, y_2) = (12, 11)$. Then

$$\begin{aligned} m &= \frac{y_2 - y_1}{x_2 - x_1} \\ &= \frac{11 - 11}{12 - 8} \\ &= \frac{0}{4} \\ &= 0, \end{aligned}$$

as desired.

25b. Use the point-slope equation, $y - y_1 = m(x - x_1)$, with one of the points, say $(x_1, y_1) = (8, 11)$:

$$\begin{aligned} y - y_1 &= m(x - x_1) \\ y - 11 &= 0(x - 8) \\ y - 11 &= 0 \\ y - 11 + 11 &= 0 + 11 \\ y &= 11 \\ &\text{or} \\ y &= 0x + 11 \end{aligned}$$

The y-intercept is $(0, 11)$.

25c. Let $(x, y) = (8, 11)$. We substitute this point into $y = 0x + 11$:

$$\begin{aligned} 11 &= 0(8) + 11 \\ &= 11 \end{aligned}$$

The point checks.

Let $(x, y) = (12, 11)$. We substitute this point into $y = 0x + 11$:

$$\begin{aligned} 11 &= 0(12) + 11 \\ &= 11 \end{aligned}$$

The point checks.

Let $(x, y) = (0, 11)$. We substitute this point into $y = 0x + 11$:

$$\begin{aligned} 11 &= 0(0) + 11 \\ &= 11 \end{aligned}$$

The point checks.

25d.

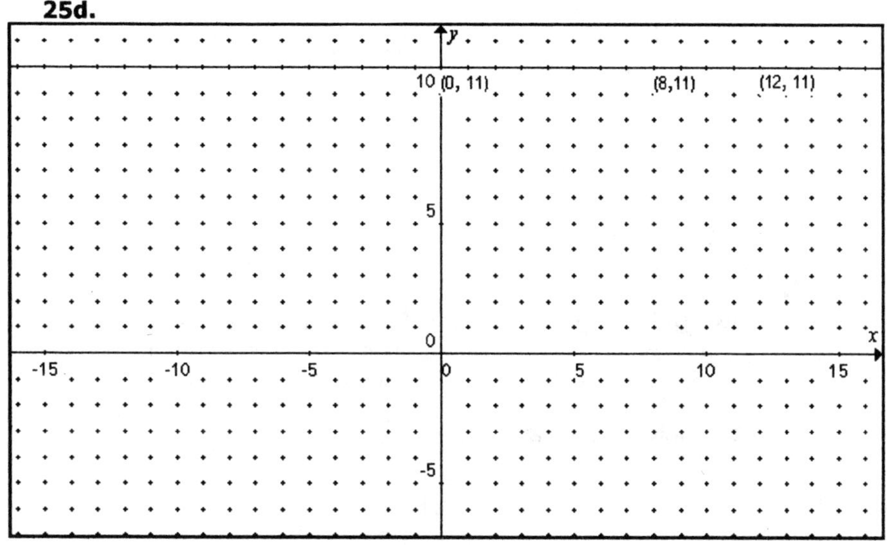

29a. $(n_1, c_1) = (1000, 900)$, $(n_2, c_2) = (800, 780)$.

29b. $m = \frac{c_2 - c_1}{n_2 - n_1} = \frac{780-900}{800-1000} = \frac{-120}{-200} = \frac{12}{20} = \frac{3}{5}$. The slope is the rate of change of cost per cone.

29c. Use the point slope equation, *mutatis mutandis:*

$$\begin{aligned} c - c_1 &= m(n - n_1) \\ c - 900 &= \frac{3}{5}(n - 1000) \\ c - 900 &= \frac{3}{5}n - \frac{3}{5} \cdot 1000 \\ c - 900 &= \frac{3}{5}n - 3 \cdot 200 \\ c - 900 &= \frac{3}{5}n - 600 \\ c - 900 + 900 &= \frac{3}{5}n - 600 + 900 \\ c &= \frac{3}{5}n + 300 \end{aligned}$$

29d.

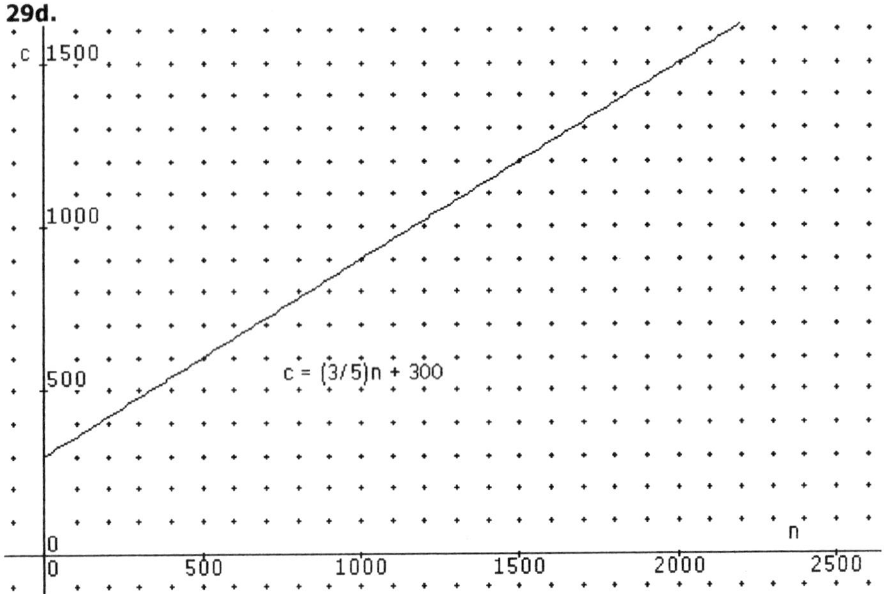

29e. The y-intercept is $(0, 300)$. The intercept represents "fixed" costs, which are the costs of operating the business itself regardless of how many cones are sold. These costs typically include rent and maintenance.

33a. Solve the equation for y:

$$
\begin{aligned}
3x + y &= 9 \\
3x - 3x + y &= -3x + 9 \\
y &= -3x + 9
\end{aligned}
$$

33b. Since the equation is in slope-intercept form, we can read the slope and y-intercept directly from the equation: $m = -3$, y-intercept $= 9$.

33c.

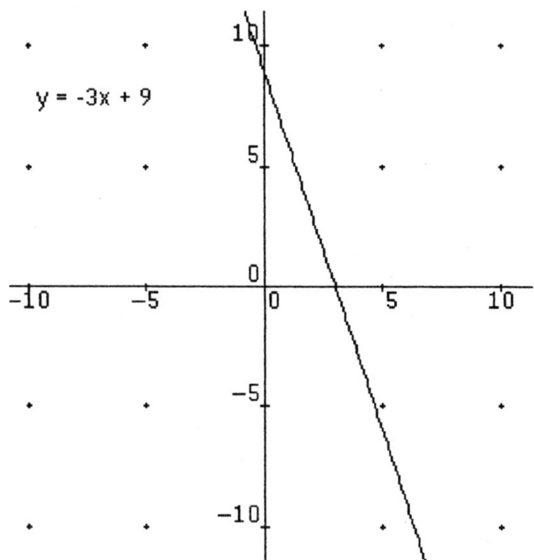

37a.

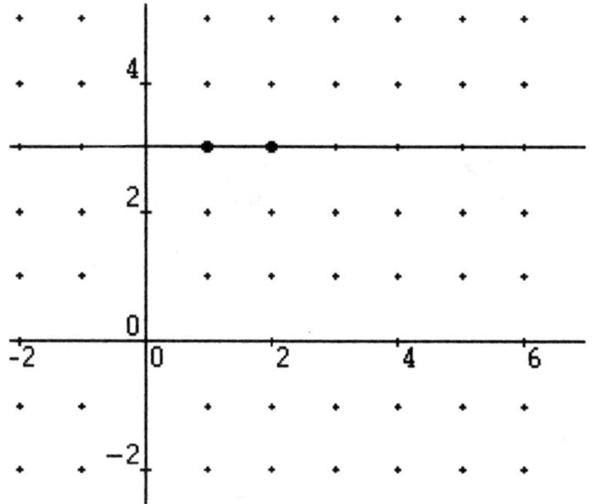

37b. Let $(x_1, y_1) = (1, 3)$, $(x_2, y_2) = (2, 3)$.

$$\begin{aligned} m &= \frac{y_2 - y_1}{x_2 - x_1} \\ &= \frac{3 - 3}{2 - 1} \\ &= \frac{0}{1} \\ &= 0 \end{aligned}$$

37c. We use the point-slope equation $y - y_1 = m(x - x_1)$ with the point $(x_1, y_1) = (1, 3)$:

$$\begin{aligned} y - y_1 &= m(x - x_1) \\ y - 3 &= 0(x - 1) \\ y - 3 &= 0 \\ y - 3 + 3 &= 0 + 3 \\ y &= 3 \end{aligned}$$

37d. $(0, 3)$, $(\pi, 3)$, $(-125, 967, 3)$ (answers may vary)

37e. We could have found the equation of the line without using the point-slope equation by noting that the y-coordinate of each point given is the constant 3.

41a. Note that the x-coordinates in both given points is -3; this means that the line containing these points is vertical, and has no slope.

Algebraically, let $(x_1, y_1) = (-3, 1)$, $(x_2, y_2) = (-3, 2)$.

$$\begin{aligned} m &= \frac{y_2 - y_1}{x_2 - x_1} \\ &= \frac{2 - 1}{-3 - (-3)} \\ &= \frac{1}{-3 + 3} \\ &= \frac{1}{0}, \end{aligned}$$

which is undefined; in other words, the slope does not exist. Note that a nonexistent slope is different from a zero slope because zero is a number; it exists. $\frac{1}{0}$ is not a number and is undefined. Thus the slope does not exist.

41b. Since both x-coordinates in the given points are -3, the line is $x = -3$.

45a. (Your answer may vary.)

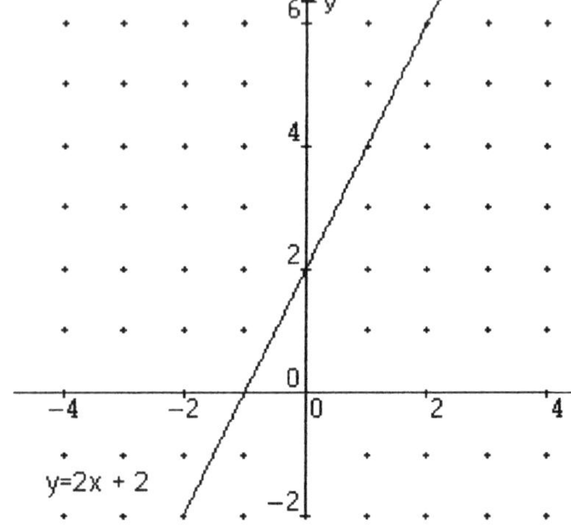

45b. (Your answer may vary.)

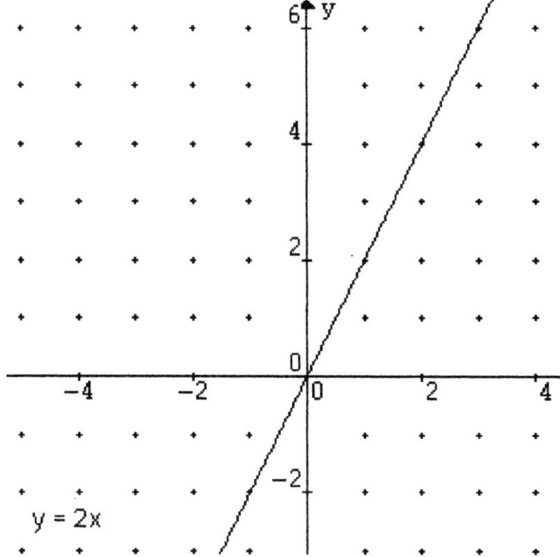

45c. (Your answer may vary.)

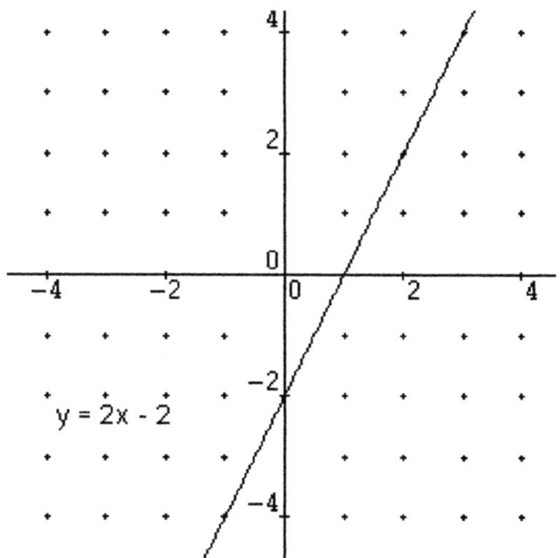

45d. (Your answer may vary.)

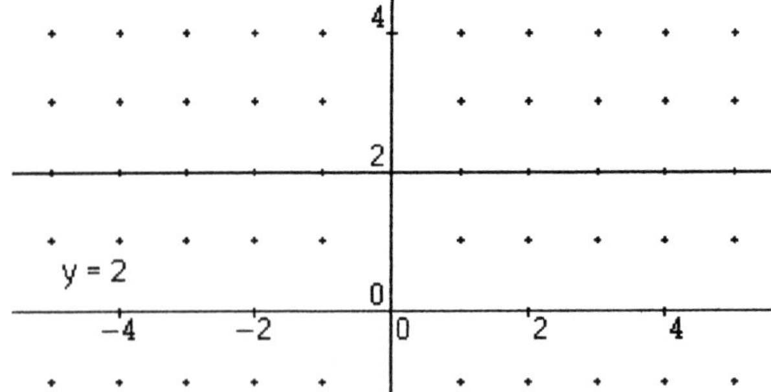

45e.

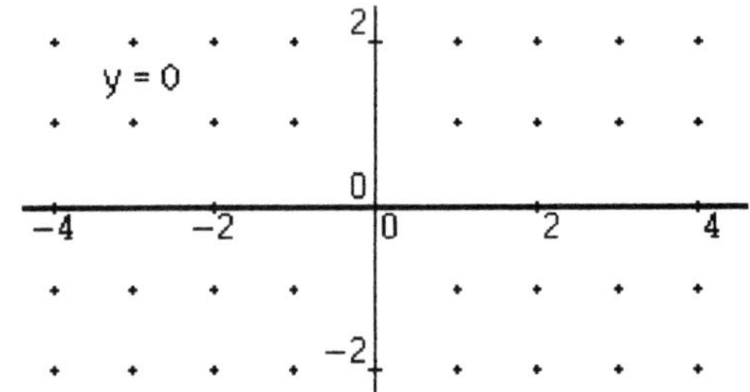

45f. (Your answer may vary.)

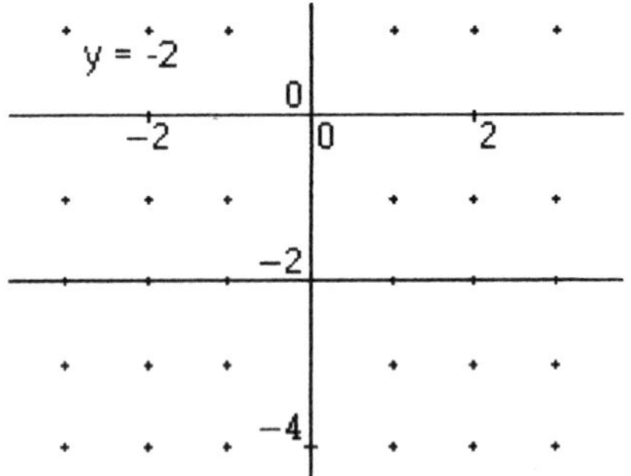

45g. (Your answer may vary.)

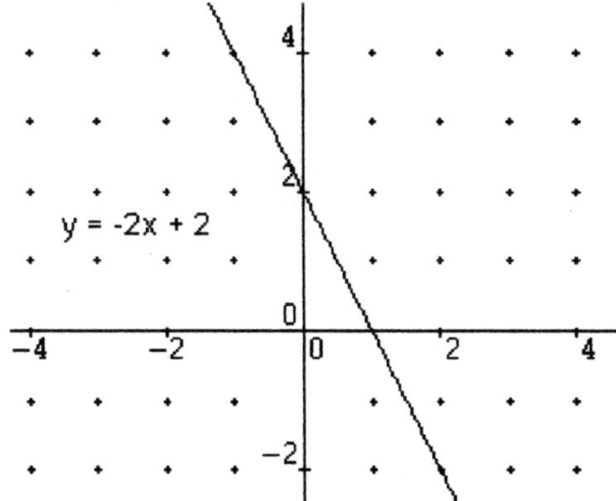

45h. (Your answer may vary.)

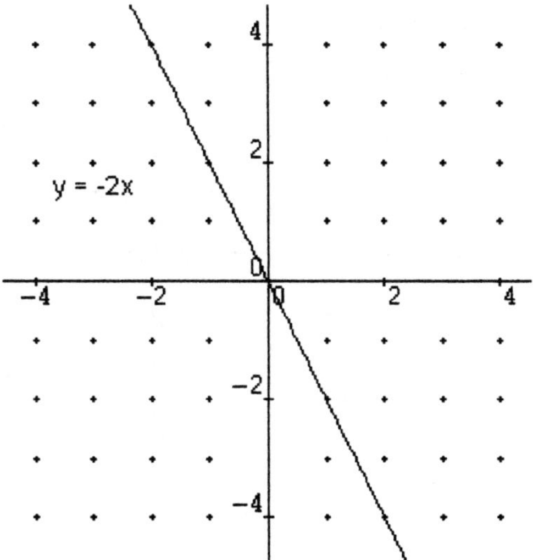

45i. (Your answer may vary.)

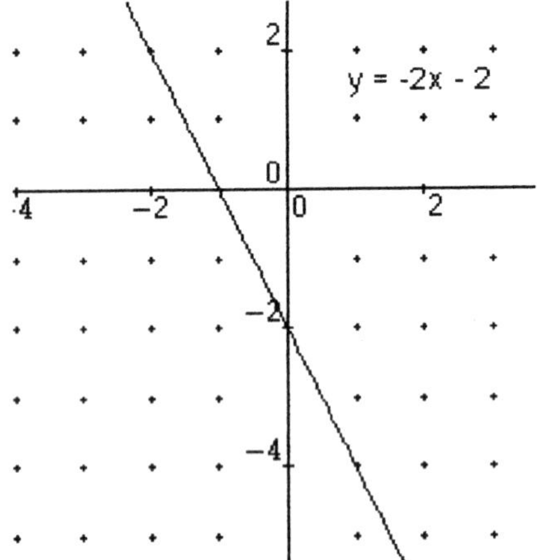

45j. (Your answer may vary.)

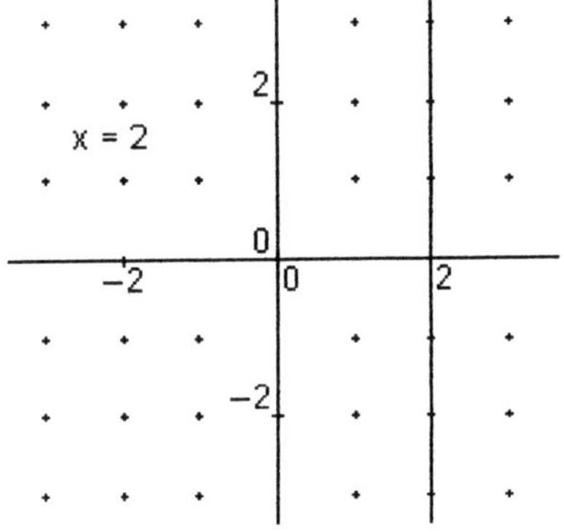

1a. $p(1000) = 310,000,000$

1b. The world's population in the year 1000 A.D. was 310,000,000.

5a. $f(3) = 2.$

5b. The value of f at $x = 3$ is 2.

9a. Yes; each state has a unique capitol.

9b. Yes; since no two capitols share the same name, then each capitol can be associated with a unique state.

13a. Yes; each instructor has a unique course.

13b. No; counterexample: *statistics* is associated with *Mr. Keating* and *Ms. Landre.*

17a. No; an input of 4 (i.e., $x = 4$) has output $y = \pm 2$.

21a. Not a function; counterexample: 1 is mapped to 2.5 and −2; i.e., we have the points (1, 2.5) and (1, −2).

25a. c; the "flat spot" at the top indicates that you were a constant distance from your home for a certain length of time, which means you were either driving in a circle around your home (not likely), or you had stopped for gas. The return home is indicated by the distance from home gradually returning to zero (where the line touches the t-axis).

25b. d; again, the flat spot indicates a stop for gas, but this time, instead of the distance decreasing, it increases, thus indicating your continuation to school.

25c. b; the steeper line segment indicates a higher speed than that indicated by the other segment. That is, more distance is travelled over the same period of time in the steeper segment than in the other segment. That probably means a drive followed by a walk.

25d. a, for reasons similar to 25c above.

29. If $0 \le x \le 10$, then

$$\begin{aligned} 0 \le 2x &\le 2 \cdot 10 \\ 0 \le 2x &\le 20 \\ 0 + 4 \le 2x + 4 &\le 20 + 4 \\ 4 \le 2x + 4 &\le 24 \\ 4 \le f(x) &\le 24. \end{aligned}$$

That is, the range is $\{y | 4 \le y \le 24\}$.

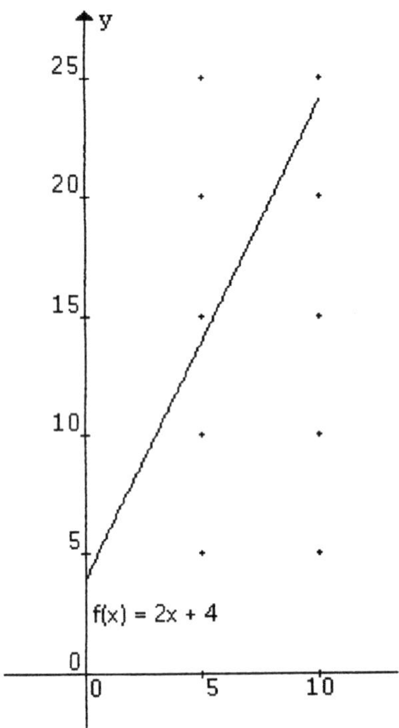

33a.

$$\begin{aligned} h(-4) &= (-4)^2 \\ &= (-1 \cdot 4)^2 \\ &= (-1)^2(4)^2 \\ &= 1 \cdot 16 \\ &= 16 \end{aligned}$$

33b.

$$\begin{aligned} -h(4) &= -(4)^2 \\ &= -1 \cdot (4)^2 \\ &= -1 \cdot 16 \\ &= -16 \end{aligned}$$

33c. $-h(-4) = -16$

37a.

$$\begin{aligned} f(g(x)) &= f(3x-1) \\ &= 2(3x-1) \\ &= 6x-2 \end{aligned}$$

37b.

$$\begin{aligned} g(f(x)) &= g(2x) \\ &= 3(2x) - 1 \\ &= 6x - 1 \end{aligned}$$

41a. The equation $c = \frac{3}{5}n + 300$ is a function because every input n has a unique output c equalling $\frac{3}{5}n + 300$.

41b. $C(n) = \frac{3}{5}n + 300$

45a. $(d_1, p_1) = (0, 14.7)$, $(d_2, p_2) = (33, 29.4)$

45b. Slope $= m = \frac{p_2 - p_1}{d_2 - d_1} = \frac{29.4 - 14.7}{33 - 0} = \frac{14.7}{33} = \frac{147}{330} = \frac{49}{110}$; for every increase in depth of 49 feet, the pressure increases 110 lbs./in.^{2}

45c.

$$\begin{aligned} p - p_1 &= m(d - d_1) \\ p - 14.7 &= \frac{49}{110}(d - 0) \\ p - 14.7 &= \frac{49}{110}d \\ p - 14.7 + 14.7 &= \frac{49}{110}d + 14.7 \\ p &= \frac{49}{110}d + 14.7 \end{aligned}$$

45d. Just write $f(d)$ (f as a function of d) instead of y in **35c**: $f(d) = \frac{49}{110}d + 14.7.$

45e. $\{d | d \geq 0\}$

$$\begin{aligned} d &\geq 0 \\ \frac{49}{110}d &\geq \frac{49}{110} \cdot 0 \\ \frac{49}{110}d &\geq 0 \\ \frac{49}{110}d + 14.7 &\geq 14.7 \\ p &\geq 14.7 \end{aligned}$$

$\{p | p \geq 14.7\}$

Actually, the domain $= \{d | 0 \leq d \leq 36,201\}$. (The upper bound of 36,201 feet is the deepest known point in the world, located in the Marianas Trench, as recorded by Soviet scientists in 1959.) Then range $= \{p | 14.7 \leq p \leq 16,140.6\}$.

45f. $f(60) = \frac{49}{110}(60) + 14.7 \approx 41.43$ lbs./in.^{2}; about 2.8 times the pressure at sea level.

45g. $f(100) = \frac{49}{110}(100) + 14.7 \approx 59.25$ lbs./in.^{2}; about 4 times the pressure at sea level.

45h. $f(12,540) = \frac{49}{110}(12,540) + 14.7 \approx 5601$ lbs./in.^{2}; 381 times the pressure at sea level.

45i.

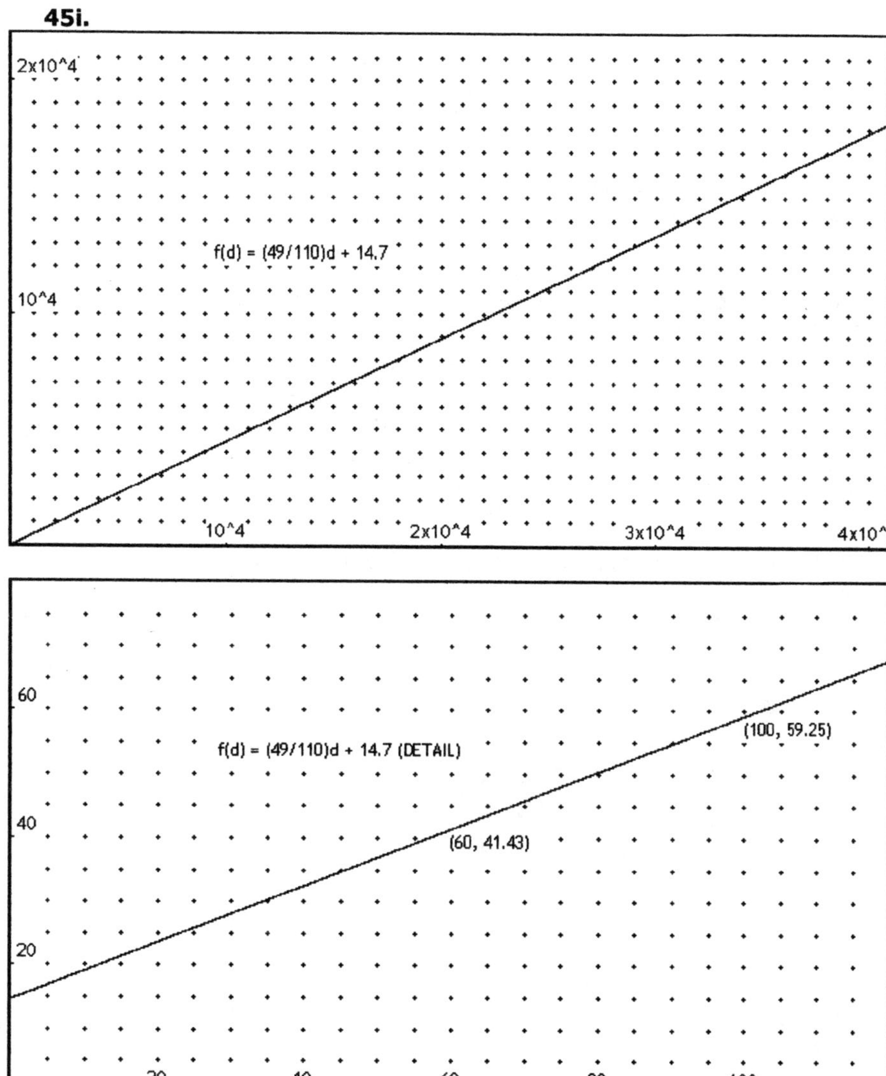

49a. When it says that there is an average global temperature rise of 0.3° C. every 10 years, it is giving us the slope of a linear function. That is, the rise is 0.3° and the run is 10, whence the slope is $m = \frac{0.3}{10} = 0.03$. Also, in 1970 the average global temperature was 15° C, and since we want a linear function with input the number of years after 1970 and with output the (predicted) average global temperature, we have the point $(y_1, t_1) = (0, 15)$, where y is years after 1970 and t is average global temperature (°C). Thus we can use the point-slope equation to derive the linear function we seek:

$$\begin{aligned} t - t_1 &= m(y - y_1) \\ t - 15 &= 0.03(y - 0) \\ t - 15 &= 0.03y \\ t &= 0.03y + 15 \end{aligned}$$

49b. The Panel's information must result in a linear function because the predicted temperature increase over time is a constant.

49c. The rise in temperature would cause melting of polar ice, thus causing a rise in sea level. To see when this would happen, let $t = 19$ in the equation in 39a and solve for y:

$$\begin{aligned} 19 &= 0.03y + 15 \\ 19 - 15 &= 0.03y \\ 4 &= 0.03y \\ \frac{4}{0.03} &= \frac{0.03y}{0.03} \\ 133.\overline{33} &= y \end{aligned}$$

Thus this would occur in $1970 + 133.\overline{33}$, or a third of the way through the year 2103.

Section 1.2 Linear Models in Business and Economics

1a. In Example 8 it was shown that \$9.50 is the equilibrium price, therefore we can use either the supply function $S(p) = 100p - 200$ or the demand function $D(p) = -500p + 5,500$ to determine the number of shirts the Business Club ordered; we use the former:

$$\begin{aligned} S(9.50) &= 100(9.50) - 200 \\ &= 950 - 200 \\ &= 750 \end{aligned}$$

They ordered 750 T-shirts.

1b. Revenue is simply the price, \$9.50 in this case, times the number x of shirts sold: $R(x) = 9.5x$. Since 750 T-shirts were sold, the revenue was $9.5(750) = \$7125.00$.

1c.

$$\begin{aligned} &C(x) = 4.50x + 50 \\ &C(750) = 4.50(750) + 50 \\ &= 3375 + 50 \\ &= 3425 \end{aligned}$$

1d. The club's profit was revenue minus cost, or \$7125 – \$3425 = \$3700.

5a. Fixed costs (do not vary with the level of production): managers' wages and fire insurance; variable costs (change with the level of production): metal tubing, bearings, assembly workers' wages (overtime), and workman's compensation insurance (premiums probably depend on production levels/hours of work).

5b. Cost function = variable costs + fixed cost = $C(x) = 57.19x + 125,418$, where x is the number of bikes produced; revenue = price per bike times number of bikes = $R(x) = 118.53x$;
profit = revenue – cost = $P(x) = R(x) - C(x) = 118.53x - (57.19x + 125,418) = 61.34x - 125,418$. The domain of each function is the set of all nonnegative integers, since it consists of numbers of bikes (although for convenience we will use the set of all nonnegative real numbers). The range for $C(x)$is $\{y|y \geq 125,418\}$, for $R(x)$is $\{y|y \geq 0\}$, and for $P(x)$is $\{y|y \geq -125,418\}$.

5c. *Hint:* Solve $R(x) = C(x)$ for x.

5d.

$$\begin{aligned} C(2556) &= 57.19(2556) + 125,418 \\ &= 271,595.64; \\ R(2556) &= 118.53(2556) \\ &= 302,962.68; \\ P(2556) &= 61.34(2556) - 125,418 \\ &= 31,367.04. \end{aligned}$$

5e.

$$\begin{aligned} C(1534) &= 57.19(1534) + 125{,}418 \\ &= 213{,}147.46; \\ R(1534) &= 118.53(1534) \\ &= 181{,}825.02; \\ P(1534) &= 61.34(1534) - 125{,}418 \\ &= -31{,}322.44. \end{aligned}$$

5f.

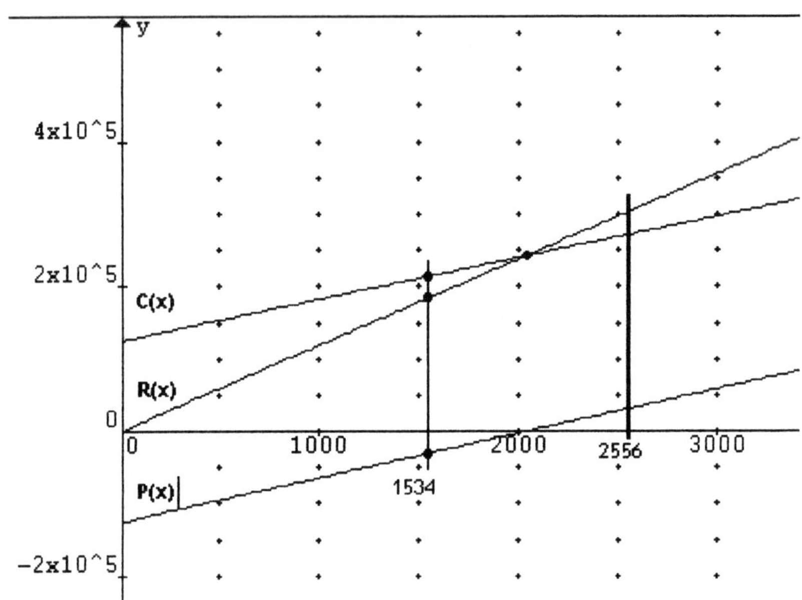

5g. Marginal cost:

$$\begin{aligned} C(x+1) - C(x) &= 57.19(x+1) + 125{,}418 - (57.19x + 125{,}418) \\ &= 57.19x + 57.19 + 125{,}418 - 57.19x - 125{,}418 \\ &= 57.19 \end{aligned}$$

Marginal revenue:

$$\begin{aligned} R(x+1) - R(x) &= 118.53(x+1) - 118.53x \\ &= 118.53x + 118.53 - 118.53x \\ &= 118.53 \end{aligned}$$

Marginal profit:

$$\begin{aligned} P(x+1) - P(x) &= 61.34(x+1) - 125{,}418 - (61.34x - 125{,}418) \\ &= 61.34x + 61.34 - 125{,}418 - 61.34x + 125{,}418 \\ &= 61.34 \end{aligned}$$

Production increases are always justified.

9a. Since $0.63(648) = 408.24$ and $0.34(648) = 220.32$ (in millions of gallons), we have the points $(p_1,\ d_1) = (5,\ 408.24)$ and $(p_2,\ d_2) = (10,\ 220.32)$, where the first coordinates are cents.

9b. We use the formula developed from Sec. 1.0, Exercise 19a: $y - y_1 = \frac{y_2 - y_1}{x_2 - x_1}(x - x_1)$. In this problem, it becomes $d - d_1 = \frac{d_2 - d_1}{p_2 - p_1}(p - p_1)$, or $D(p) = \frac{d_2 - d_1}{p_2 - p_1}(p - p_1) + d_1$.

$$\begin{aligned} D(p) &= \frac{d_2 - d_1}{p_2 - p_1}(p - p_1) + d_1 \\ &= \frac{220.32 - 408.24}{10 - 5}(p - 5) + 408.24 \\ &= \frac{-187.92}{5}(p - 5) + 408.24 \\ &= -37.584p + 187.92 + 408.24 \\ &= -37.584p + 596.16 \end{aligned}$$

9c. Domain:

$$\begin{aligned} D(p) &\geq 0 \\ -37.584p + 596.16 &\geq 0 \\ 37.584p &\leq 596.16 \\ p &\leq 15.862, \end{aligned}$$

range :

$$\begin{aligned} p &> 0 \\ -37.584p &< 0 \\ -37.584p + 596.16 &< 596.16 \\ D(p) &< 596.16, \end{aligned}$$

that is, the domain is $(0, 15.862]$ and the range is $[0,\ 596.16)$.

9d.

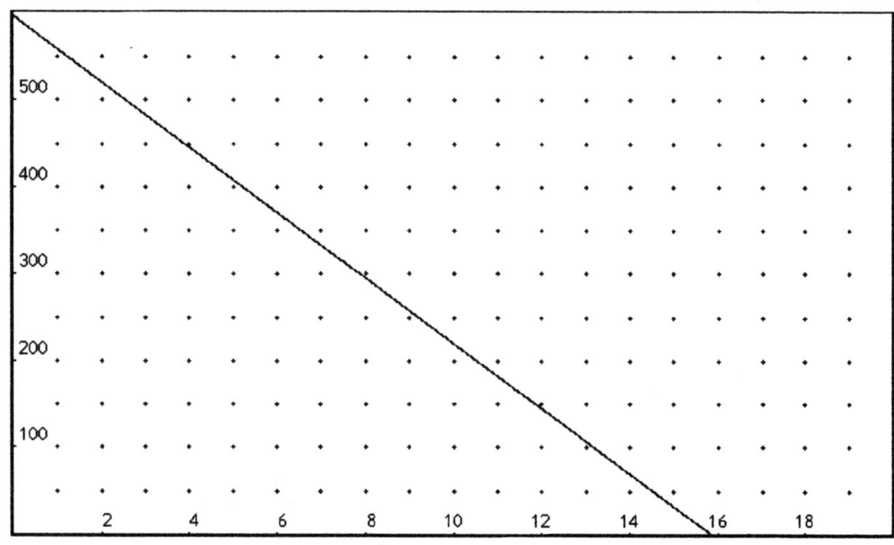

13a. We have the points $(t_1,\ v_1) = (0, 13, 500)$, $(t_2,\ v_2) = (7,\ 0)$, where t_n is the age of the drill press in years and v_n is the value of the drill press in dollars. We use the formula developed from Sec. 1.0, Exercise 19a: $y - y_1 = \frac{y_2 - y_1}{x_2 - x_1}(x - x_1)$.

In this problem, it becomes $v - v_1 = \frac{v_2 - v_1}{t_2 - t_1}(t - t_1)$, or $V(t) = \frac{v_2 - v_1}{t_2 - t_1}(t - t_1) + v_1$.

$$\begin{aligned} V(t) &= \frac{v_2 - v_1}{t_2 - t_1}(t - t_1) + v_1 \\ &= \frac{0 - 13,500}{7 - 0}(t - 0) + 13,500 \\ &= \frac{-13,500}{7}t + 13,500 \\ &\approx -1928.57t + 13,500 \end{aligned}$$

13b. $V(1) = -1928.57(1) + 13,500 = 11,571.43.$

13c. $V(2) = -1928.57(2) + 13,500 = -3857.14 + 13,500 = 9642.86.$

13d. This can be seen directly from the slope; about \$1928.57 of value is lost each year.

13e. \$1928.57

17. Take the formula for the demand function from Example 6, $D(p) = -500p + 5500$, and solve it for p:

$$\begin{aligned} D(p) &= -500p + 5500 \\ 500p &= -D(p) + 5500 \\ p &= -\frac{D(p)}{500} + \frac{5500}{500} \\ &= -\frac{D(p)}{500} + 11 \end{aligned}$$

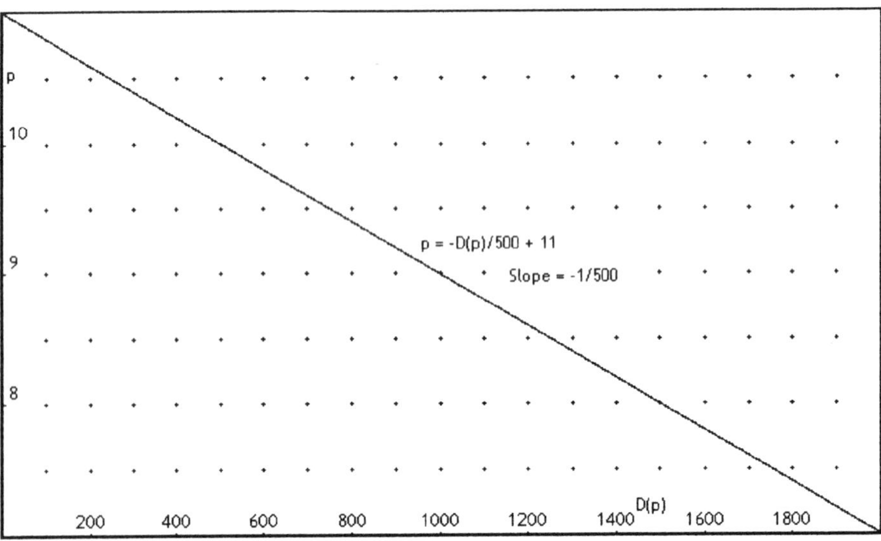

21. A break-even point occurs where the company concerned will have no profit nor loss; they are making enough, and only enough, revenue to cover operational costs. Other points are either indicative of a profit situation, where the revenue is greater than the costs and the company is making a profit, or a loss situation, where the operational costs are greater than the revenue and the company is losing money, or is operating "in the red."

The demand is the quantity of a product which consumers will purchase at a given price; think of it as "offers to buy" at a street market. The supply is the quantity of a product that sellers are willing to offer at a given price; think of it as "offers to sell" at a street market. At a high price a seller would be willing to offer many items for sale, but he or she will attract few buyers because the demand is low. At a low price the opposite holds. The equlibrium price is the price at which offers to buy are equal to offers to sell; the market is at equilibrium. This occurs where the supply and demand curves intersect.

1e. The slope of the line of best fit is:

$$\begin{aligned} m &= \frac{n(\sum xy) - (\sum x)(\sum y)}{n(\sum x^2) - (\sum x)^2} \\ &= \frac{3(1 \cdot 10 + 5 \cdot 4 + 9 \cdot 1) - (1+5+9)(10+4+1)}{3(1^2+5^2+9^2) - (1+5+9)^2} \\ &= \frac{3(39) - (15)(15)}{3(107) - 15^2} \\ &= \frac{117 - 225}{321 - 225} \\ &= \frac{-108}{96} \\ &= -1.125, \end{aligned}$$

and using some of the above results, the y-intercept is:

$$\begin{aligned} b &= \frac{(\sum y) - m(\sum x)}{n} \\ &= \frac{15 - (-1.125)(15)}{3} \\ &= \frac{15 + (1.125)(15)}{3} \\ &= \frac{15(1 + 1.125)}{3} \\ &= 5(1 + 1.125) \\ &= 5(2.125) \\ &= 10.625. \end{aligned}$$

Therefore the line of best fit is $\widehat{y} = -1.125x + 10.625$.

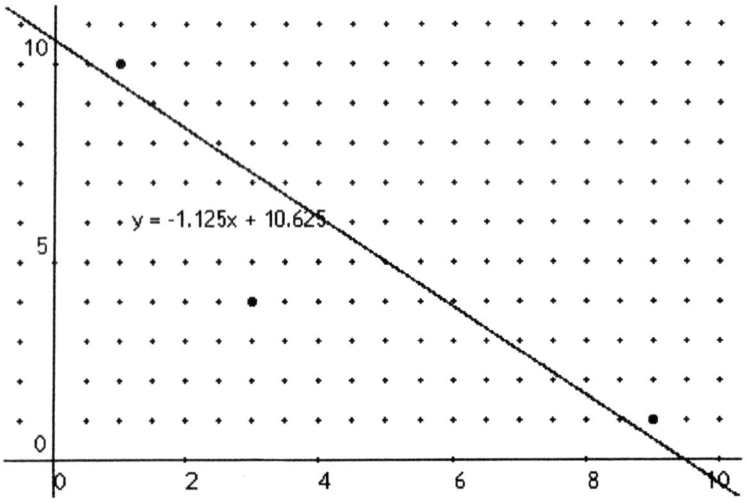

1f. The error at $x = 1$ is $10 - 9.5 = 0.5$, and the relative error is then $\frac{0.5}{10} = 0.05$.

5a. Relative error at $x = 1$ is $0.082/1 = 0.082$, at $x = 5$ is $0.146/3 = 0.04867$, and at $x = 10$ is $0.069/5 = 0.0138$.

5b. Correlation coefficient r:

$$\begin{aligned}
r &= \frac{n(\sum xy) - (\sum x)(\sum y)}{\sqrt{n(\sum x^2) - (\sum x)^2}\sqrt{n(\sum y^2) - (\sum y)^2}} \\
&= \frac{3(1 \cdot 1 + 5 \cdot 3 + 10 \cdot 5) - (1 + 5 + 10)(1 + 3 + 5)}{\sqrt{3(1^2 + 5^2 + 10^2) - (1 + 5 + 10)^2}\sqrt{3(1^2 + 3^2 + 5^2) - (1 + 3 + 5)^2}} \\
&= \frac{3(66) - (16)(9)}{\sqrt{3(1 + 25 + 100) - 16^2}\sqrt{3(1 + 9 + 25) - 9^2}} \\
&= \frac{54}{\sqrt{122}\sqrt{24}} \\
&= \frac{54}{\sqrt{122}\sqrt{4 \cdot 6}} \\
&= \frac{54}{2\sqrt{122}\sqrt{6}} \\
&= \frac{27}{\sqrt{122 \cdot 6}} \\
&= \frac{27}{\sqrt{732}} \\
&\approx 0.99794871579
\end{aligned}$$

9a. (11, 255.4), (23, 250.4), (33, 247.6), (42, 244.6), (45, 241.4), (54, 239.4), (64, 234.1), (67, 231.1), (75, 229.4), (80, 228.8), (85, 226.3)

9b. Slope: -0.403423262518; y-intercept: 260.189279; correlation coefficient: -0.994913366659.

9c. $\hat{y} = -0.403423262518x + 260.189279$

9d. $-0.403423262518(100) + 260.189279 \approx 219.8$ sec., or about $3 : 39.5$ (min:sec).

9e. $3\frac{1}{2}$ min. = 210 sec.; solve the equation $210 = -0.403423262518x + 260.189279$ and get $x \approx 124.4$, or the record time for the mile run will reach 3.5 min. in the year 2024.

9f. The assumption is that past trends in the mile run reliably predict future trends, and if this assumption is correct, the predictions are highly accurate given the correlation coefficient, which is very close to -1.

13a-c. The equation of the line of best fit is

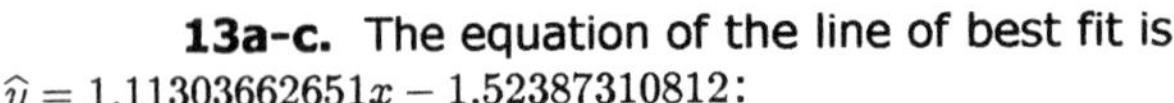

$\widehat{y} = 1.11303662651x - 1.52387310812$:

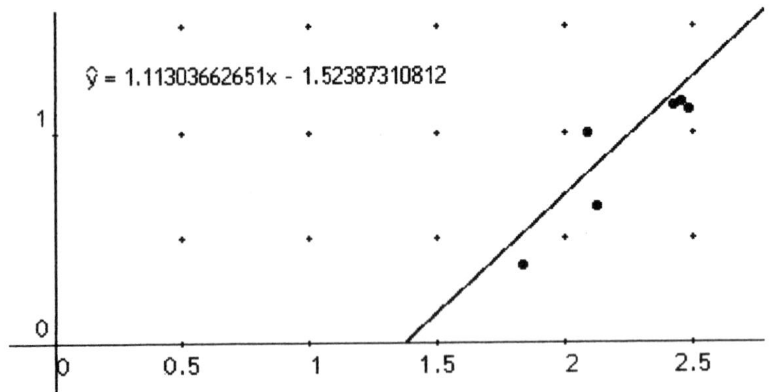

13d. $\widehat{y} = 1.11303662651(2.75) - 1.52387310812 \approx 1.5370$, or about 1,537,000 divorces.

13e. Solve $2.75 = 1.11303662651x - 1.52387310812$ for x: $x = \frac{2.75+1.52387310812}{1.11303662651} \approx 3.840$, or about 3,840,000 marriages.

13f. Correlation coefficient equals 0.933894673375, a "good fit." A positive correlation means that an increase in the independent variable results in an increase in the dependent variable.

13g. The assumption is that past trends reliably predict future trends, and if this assumption is correct, the predictions are highly accurate given the correlation coefficient, which is very close to 1.

Review Exercises

1a. Let $(x_1, y_1) = (1, 7)$, $(x_2, y_2) = (2, -1)$, and $m =$ slope.

$$\begin{aligned} m &= \frac{y_2 - y_1}{x_2 - x_1} \\ &= \frac{-1-7}{2-1} \\ &= \frac{-8}{1} \\ &= -8 \end{aligned}$$

1b.

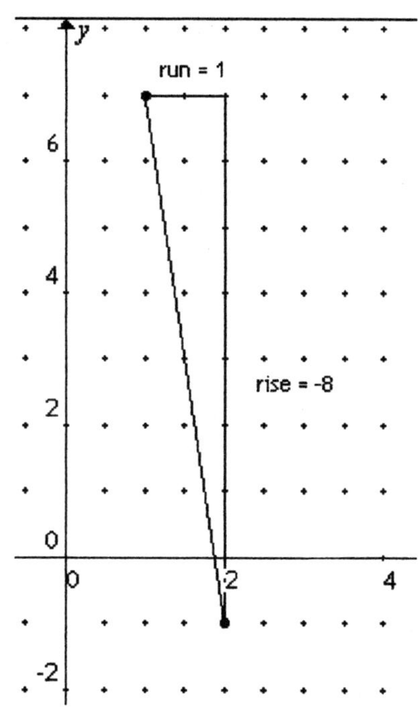

2a. Let $(x_1, y_1) = (3, -2)$, $(x_2, y_2) = (5, -2)$, and $m =$ slope.

$$\begin{aligned} m &= \frac{y_2 - y_1}{x_2 - x_1} \\ &= \frac{-2-(-2)}{5-3} \\ &= \frac{-2+2}{2} \\ &= \frac{0}{2} \\ &= 0 \end{aligned}$$

2b.

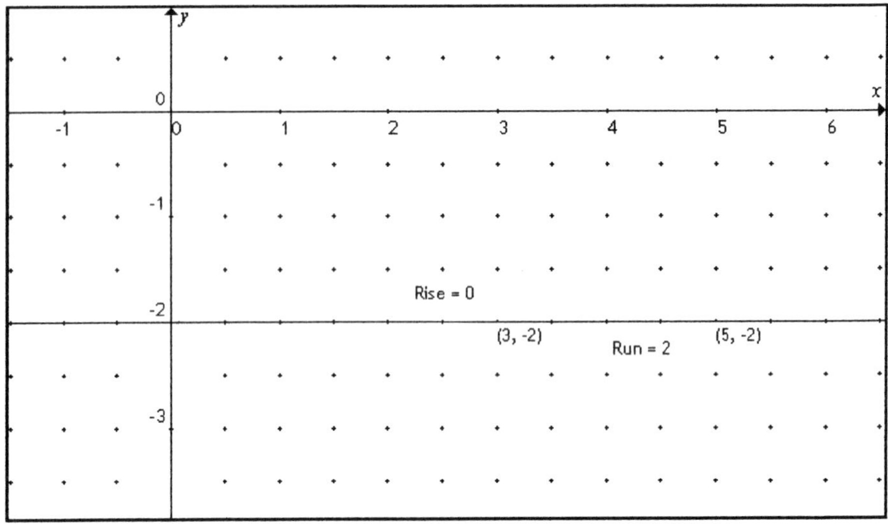

3a. Let $(x_1, y_1) = (2,\ 6)$, $(x_2, y_2) = (2,\ -1)$, and $m =$ slope.

$$
\begin{aligned}
m &= \frac{y_2 - y_1}{x_2 - x_1} \\
&= \frac{-1-6}{2-2} \\
&= \frac{-5}{0} \\
&\emptyset
\end{aligned}
$$

The slope does not exist.

3b.

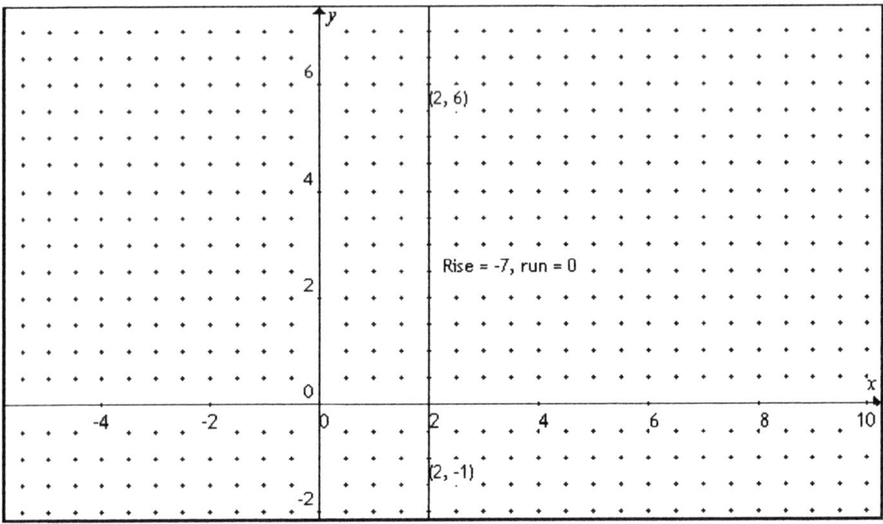

4a. Let $(x_1, y_1) = (-1, 5)$, $(x_2, y_2) = (3, 12)$, and $m =$ slope.

$$\begin{aligned} m &= \frac{y_2 - y_1}{x_2 - x_1} \\ &= \frac{12 - 5}{3 - (-1)} \\ &= \frac{7}{3 + 1} \\ &= \frac{7}{4} \end{aligned}$$

4b.

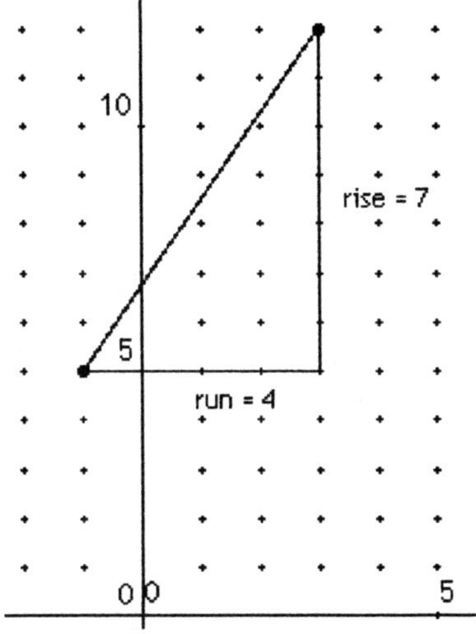

5.

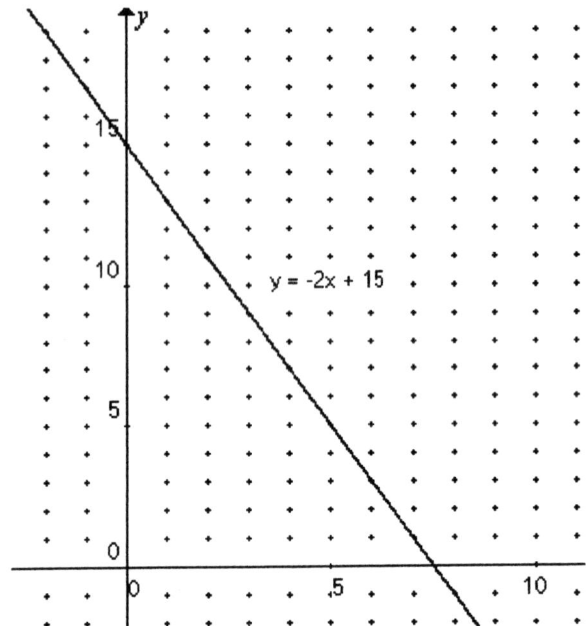

6.

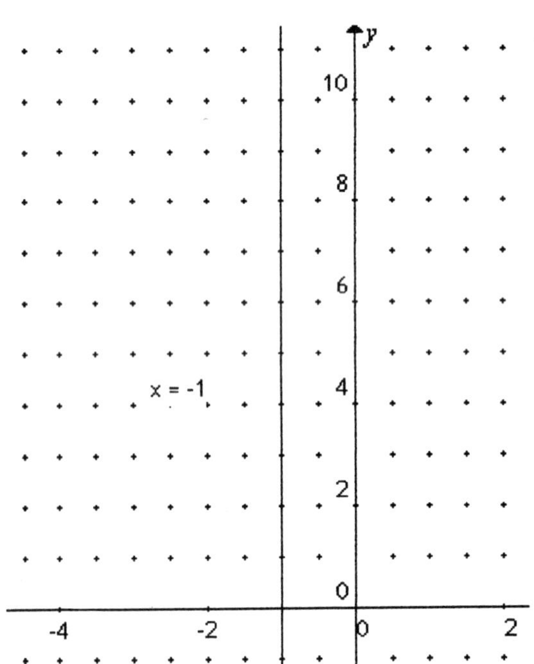

7.

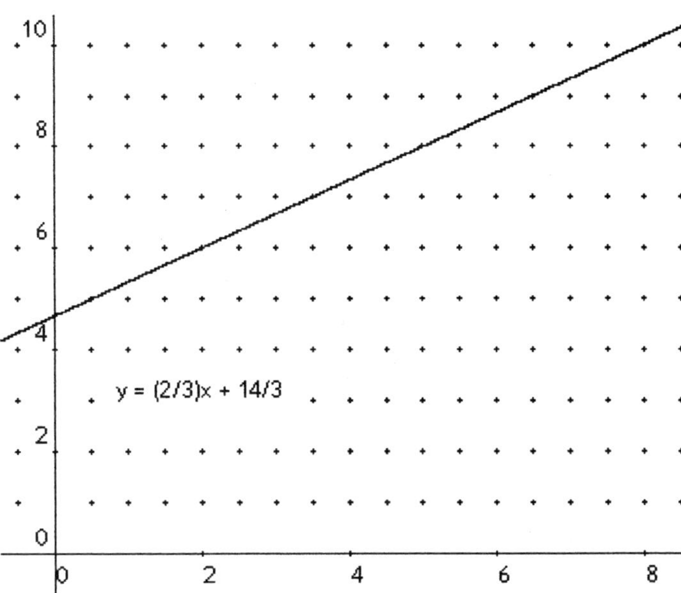

8.

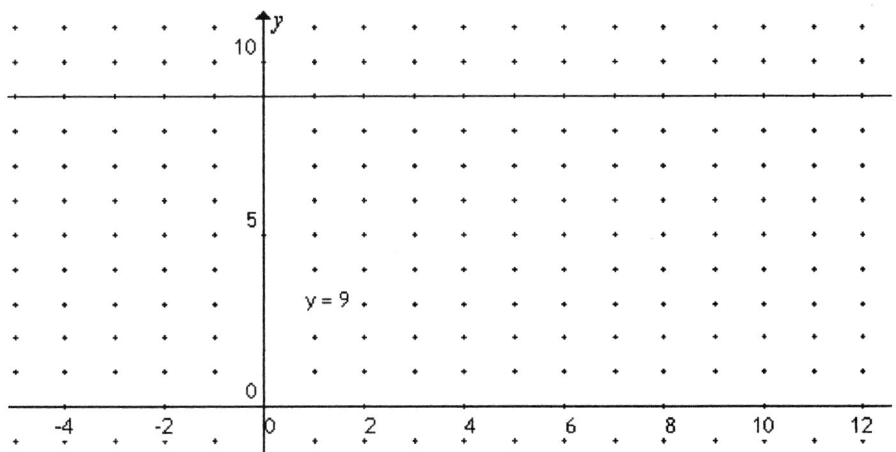

9a.

$$\begin{aligned} 2x - 6y &= 9 \\ 6y &= 2x - 9 \\ y &= \frac{2}{6}x - \frac{9}{6} \\ &= \frac{1}{3}x - \frac{3}{2} \end{aligned}$$

9b. Slope $= 1/3$, y-intercept $= -3/2$

9c.

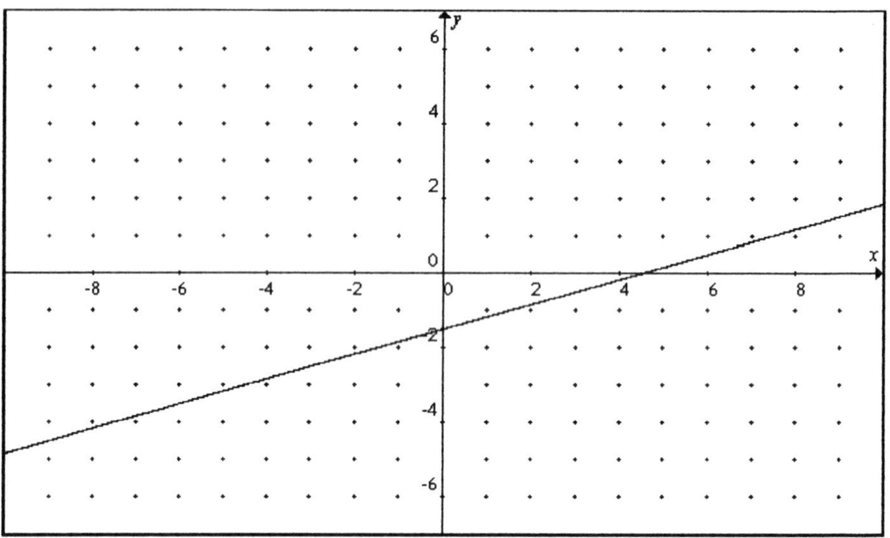

10a.

$$\begin{aligned} 4x + 2y &= 12 \\ 2y &= -4x + 12 \\ y &= \frac{-4}{2}x + \frac{12}{2} \\ &= -2x + 6 \end{aligned}$$

10b. Slope $= -2$, y-intercept $= 6$

10c.

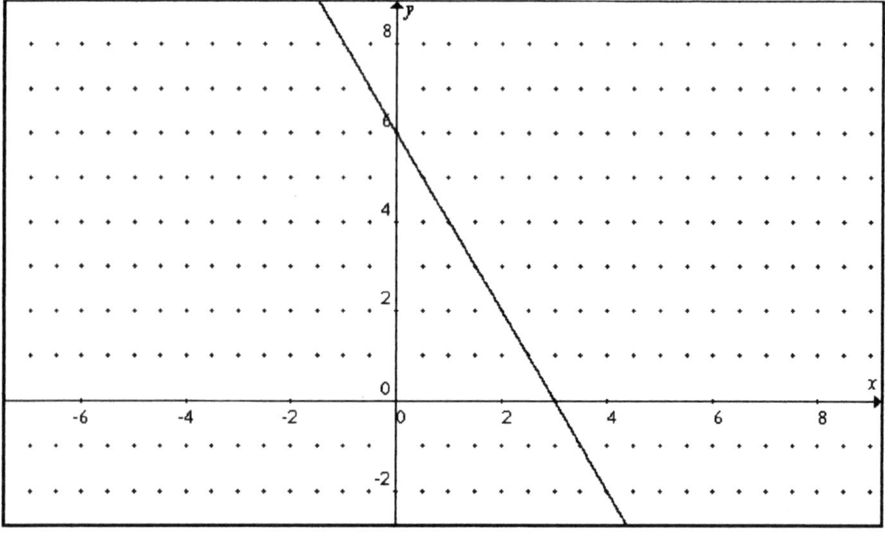

11a. Since no class can have more than one enrollment level, this relation is a function.

11b. Let $E(x)$ represent class enrollment as a function of $x =$ class.

12. This relation is not a function, because one enrollment level may apply to more than one class.

13a. This is a function; its graph is a straight line.

13b. Domain and range are all real numbers.

14a. This is a function; its graph is a parabola opening upward with its vertex at the origin.

14b. Domain is all real numbers; range is all nonnegative real numbers.

15a. This is not a function. Counterexample: Let $x = 4$; then $y^2 = x = 4$ and hence $y = \pm\sqrt{4} = \pm 2$. In other words, there are two outputs for every input (greater than zero).

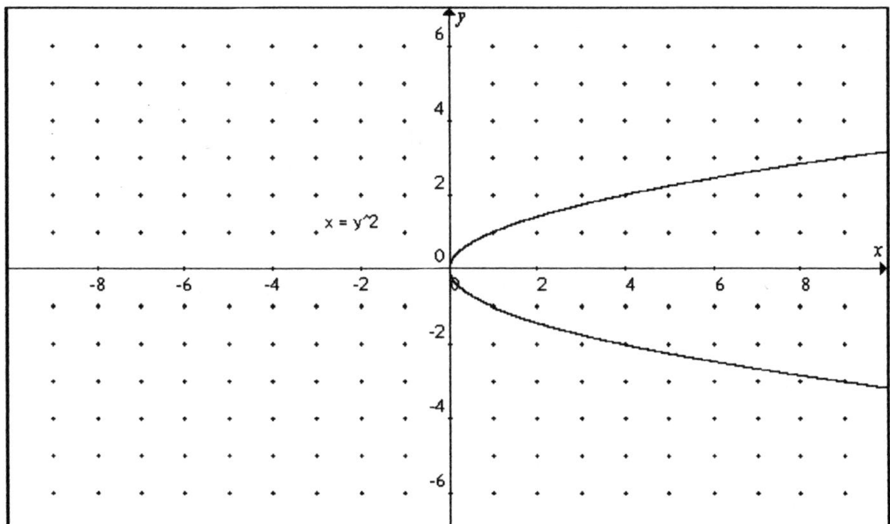

16a. This is a function because it passes the vertical line test.

16b. Domain $= \{x|-4 \le x \le 1\}$; range $= \{y|-4 \le y \le 3\}$.

17a. This is not a function because it fails the vertical line test.

18.

$$\begin{aligned} 0 &\le x \le 10 \\ 3\cdot 0 &\le 3x \le 3\cdot 10 \\ 0 &\le 3x \le 30 \\ 0-7 &\le 3x-7 \le 30-7 \\ -7 &\le 3x-7 \le 23, \end{aligned}$$

or $-7 \leq f(x) \leq 23$ (the range).

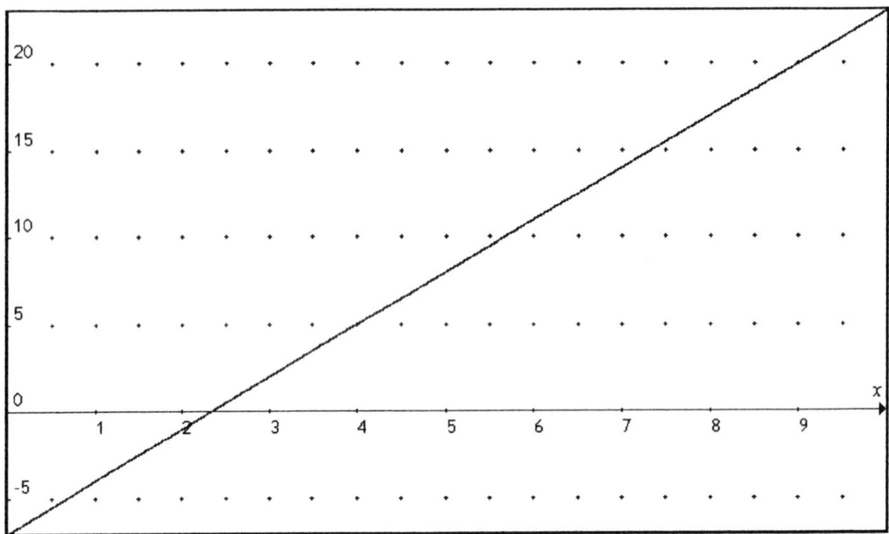

19a.

$$\begin{aligned} f(-3) &= 2(-3)+1 \\ &= -6+1 \\ &= -5 \end{aligned}$$

19b.

$$\begin{aligned} -f(3) &= -[2(3)+1] \\ &= -(6+1) \\ &= -7 \end{aligned}$$

20a.

$$g(2x-1) = (2x-1)^2$$
$$= 4x^2 - 4x + 1$$

20b.

$$g(2x) - 1 = (2x)^2 - 1$$
$$= 4x^2 - 1$$

20c.

$$2g(x) - 1 = 2\left(x^2\right) - 1$$
$$= 2x^2 - 1$$

21a.

$$\begin{aligned} f(g(x)) &= 2(g(x))+1 \\ &= 2\left(x^2\right)+1 \\ &= 2x^2+1 \end{aligned}$$

21b.

$$\begin{aligned} g(f(x)) &= [f(x)]^2 \\ &= (2x+1)^2 \\ &= 4x^2 + 4x + 1 \end{aligned}$$

22a. $(F_1,\ C_1) = (212,\ 100),\ (F_2,\ C_2) = (32,\ 0)$

22b.

$$\begin{aligned} m &= \frac{C_2 - C_1}{F_2 - F_1} \\ &= \frac{0 - 100}{32 - 212} \\ &= \frac{-100}{-180} \\ &= \frac{100}{180} \\ &= \frac{5}{9} \end{aligned}$$

Note first that the independent variable is degrees Fahrenheit. Since the slope is positive, the graph of the line rises from left to right, and for every 9° F. increase, there is a corresponding 5° C. increase.

22c.

$$\begin{aligned} C - C_1 &= m(F - F_1) \\ C - 100 &= \frac{5}{9}(F - 212) \\ C &= \frac{5}{9}F - \frac{5}{9}\cdot 212 + 100 \\ &= \frac{5}{9}F - \frac{1060}{9} + 100 \cdot \frac{9}{9} \\ &= \frac{5}{9}F + \frac{-1060 + 900}{9} \\ &= \frac{5}{9}F - \frac{160}{9} \end{aligned}$$

22d. $C(F) = \frac{5}{9}F - \frac{160}{9}$

22e.

$$\begin{aligned} C(F) &= \frac{5}{9}F - \frac{160}{9} \\ C(0) &= \frac{5}{9}\cdot 0 - \frac{160}{9} \\ &= 0 - \frac{160}{9} \\ &= -\frac{160}{9} \\ &\approx -17.78 \end{aligned}$$

22f.

$$\begin{aligned} C(98.6) &= \frac{5}{9} \cdot 98.6 - \frac{160}{9} \\ &= \frac{5 \cdot 98.6 - 160}{9} \\ &= \frac{493 - 160}{9} \\ &= \frac{333}{9} \\ &= 37 \end{aligned}$$

22g. Take the equation from 22c above and solve it for F:

$$\begin{aligned} C &= \frac{5}{9}F - \frac{160}{9} \\ C + \frac{160}{9} &= \frac{5}{9}F \\ \frac{9}{5}C + \frac{9}{5} \cdot \frac{160}{9} &= F \\ F &= \frac{9}{5}C + 32 \end{aligned}$$

22h.

$$\begin{aligned} F(C) &= \frac{9}{5}C + 32 \\ F(25) &= \frac{9}{5} \cdot 25 + 32 \\ &= 9 \cdot 5 + 32 \\ &= 77 \end{aligned}$$

23a. Fixed: managers' wages. Variable: down; cotton fabric; thread; sewing machines repair; assemblers' wages; gas and electricity.

23b. Let x represent the quantity of comforters sold.

$$\begin{aligned} \text{Cost} &= \text{variable cost} + \text{fixed cost} \\ C(x) &= 38x + 111{,}412 \end{aligned}$$

$$\begin{aligned} \text{Revenue} &= 349 \cdot \text{no. of comforters sold} \\ R(x) &= 349x \end{aligned}$$

$$\begin{aligned} \text{Profit} &= \text{revenue} - \text{cost} \\ P(x) &= 349x - (38x + 111{,}412) \\ &= 311x - 111{,}412 \end{aligned}$$

The domain for all three functions is $\{x | x \geq 0\}$, since anywhere from zero to an undetermined amount of comforters may be sold. Thus the range for $C(x)$ is $\{y | y \geq 111{,}412\}$, for $R(x)$ is $\{y | y \geq 0\}$, and for $P(x)$ is $\{y | y \geq -111{,}412\}$. (To see this, simply substitute 0 for x in each function to get the lower bounds for the ranges.)

23c. The break-even point occurs where revenue equals cost. To solve this, set $R(x) = C(x)$, solve for x, and then substitute this value of x into $R(x)$ to see the break-even revenue.

23d.

$$\begin{aligned} C(394) &= 38(394) + 111,412 \\ &= 126,384 \end{aligned}$$

$$\begin{aligned} R(394) &= 349(394) \\ &= 137,506 \end{aligned}$$

$$\begin{aligned} P(394) &= 311(394) - 111,412 \\ &= 11,122 \end{aligned}$$

23e.

$$\begin{aligned} C(322) &= 38(322) + 111,412 \\ &= 123,648 \end{aligned}$$

$$\begin{aligned} R(322) &= 349(322) \\ &= 112,378 \end{aligned}$$

$$\begin{aligned} P(322) &= 311(322) - 111,412 \\ &= -11,000 \end{aligned}$$

23f.

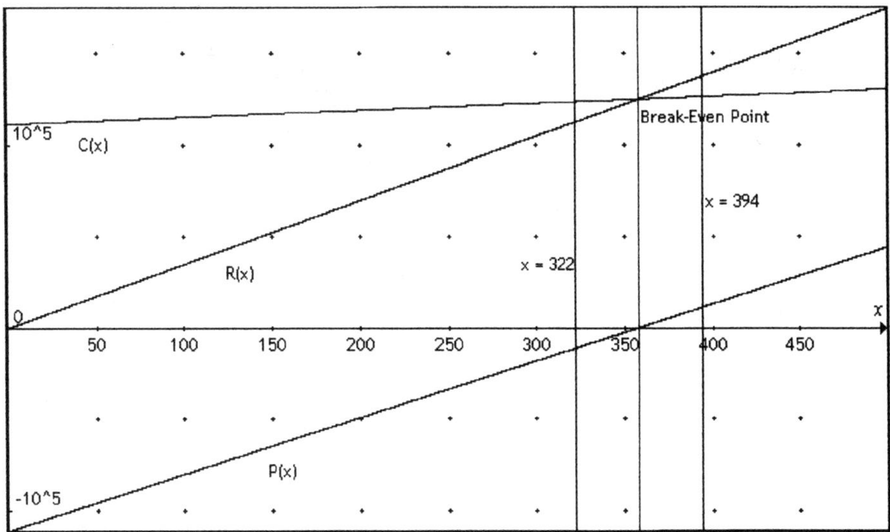

23g. Marginal revenue:

$$\begin{aligned} 349(x+1) - 349x &= 349x + 349 - 349x \\ &= 349 \end{aligned}$$

Marginal cost:

$$\begin{aligned} 38(x+1) + 111,412 - (38x + 111,412) &= 38x + 38 + 111,412 - 38x - 111,412 \\ &= 38 \end{aligned}$$

Increased production is justified since the marginal revenue exceeds the marginal cost in general.

25a. Note that we are given two data points $(t,\, v)$ where t is the number of years after purchase and v is the value in dollars: $(t_1,\, v_1) = (0,\, 7200)$, $(t_2,\, v_2) = (7,\, 0)$. Thus

$$\begin{aligned} v - v_1 &= \frac{v_2 - v_1}{t_2 - t_1}(t - t_1) \\ v - 7200 &= \frac{0 - 7200}{7 - 0}(t - 0) \\ v = v(t) &= -\frac{7200}{7}t + 7200 \end{aligned}$$

25b. $v(3) = -\frac{7200}{7} \cdot 3 + 7200 = -\frac{21{,}600}{7} + 7200 \approx 4114.29$

25c. This is seen directly from the slope; it loses $\frac{7200}{7} \approx \$1028.57$ per year.

25d. $1028.57

26b. (Let the independent variable represent the number of years since 1970.)

$$\begin{aligned} \widehat{y} &= mx + b \\ &= \frac{n(\sum xy) - (\sum x)(\sum y)}{n(\sum x^2) - (\sum x)^2}x + \frac{(\sum y) - m(\sum x)}{n} \\ &= \frac{\begin{array}{c}5(0 \cdot 12.091 + 5 \cdot 11.220 + 10 \cdot 10.758 + 15 \cdot 10.777 + 20 \cdot 10.153) - \\ (0 + 5 + 10 + 15 + 20)(12.091 + 11.220 + 10.758 + 10.777 + 10.153)\end{array}}{5(0^2 + 5^2 + 10^2 + 15^2 + 20^2) - (0 + 5 + 10 + 15 + 20)^2}x + \\ &\quad \frac{(12.091 + 11.220 + 10.758 + 10.777 + 10.153) - m(0 + 5 + 10 + 15 + 20)}{5} \\ &= \frac{5(528.395) - (50)(54.999)}{5(750) - 2500}x + \frac{54.999 - m(50)}{5} \\ &= \frac{-107.975}{1250}x + \frac{54.999 - \left(-\frac{107.975}{1250}\right)(50)}{5} \\ &= -0.08638x + \frac{54.999 + \frac{107.975}{25}}{5} \\ &= -0.08638x + 11.8636 \end{aligned}$$

26c.

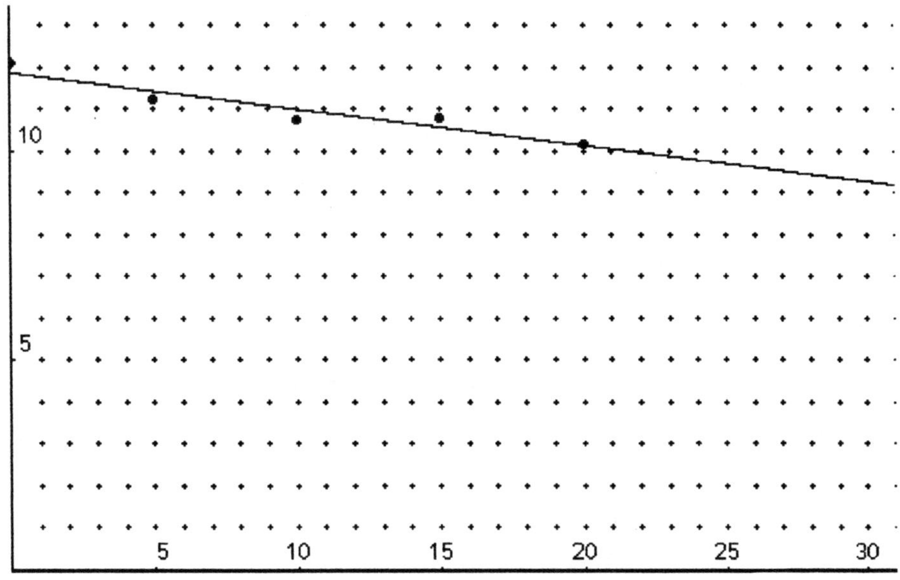

26d. Our value for x must be $2010 - 1970 = 40$. Now,

$$\begin{aligned} -0.08638(40) + 11.8636 &= -3.4552 + 11.8636 \\ &= 8.4084, \end{aligned}$$

or 8,408,400 cows.

26e. We solve $9 = -0.08638x + 11.8636$ for x:

$$\begin{aligned} 9 &= -0.08638x + 11.8636 \\ 0.08638x &= 11.8636 - 9 \\ 0.08638x &= 2.8636 \\ x &= \frac{2.8636}{0.08638} \\ &\approx 33.15, \end{aligned}$$

which means that there will be 9,000,000 cows by the year $1970 + 33.15 \approx 2003$.

26f. We use some of the data from 26b above to find the correlation coefficient r:

$$\begin{aligned} r &= \frac{n(\sum xy) - (\sum x)(\sum y)}{\sqrt{n(\sum x^2) - (\sum x)^2}\sqrt{n(\sum y^2) - (\sum y)^2}} \\ &= \frac{5(528.395) - (50)(54.999)}{\sqrt{5(750) - 2500}\sqrt{5(12.091^2 + 11.220^2 + 10.758^2 + 10.777^2 + 10.153^2) - 54.999^2}} \\ &= \frac{-107.975}{\sqrt{1250}\sqrt{5(607.042383) - 3024.890001}} \\ &\approx -\frac{107.975}{\sqrt{1250}\sqrt{10.321914}} \\ &\approx -0.950578720593 \end{aligned}$$

26g. The assumption is that past trends reliably predict future trends, and if this assumption is correct, the predictions are highly accurate given the correlation coefficient, which is very close to -1.

Chapter 2 Systems of Linear Equations and Inequalities

Section 2.0 The Elimination Method

1.

$$\begin{aligned}3(4)-5(1)&=12-5\\&=7,\\2(4)+2(1)&=8+2\\&=10\end{aligned}$$

Yes

5.

$$\begin{aligned}2(4)+3(-1)-(2)&=8-3-2\\&=3,\\(4)+(-1)+(2)&=4-1+2\\&=5\\10(4)-2(-1)&=40+2\\&=42\end{aligned}$$

No

9.

$$\begin{aligned}4x+3y&=12\\3y&=-4x+12\\y&=-\frac{4}{3}x+4\\8x+6y&=24\\6y&=-8x+24\\y&=-\frac{4}{3}x+4\end{aligned}$$

Since these are one and the same line, they intersect at all points, hence the given system has an infinite number of solutions.

13. Since none of the equations are multiples or combinations of the others, this system could have a single solution.

17. Since there are fewer equations than unknowns, this system cannot have a unique solution.

21. $(2, -3)$

25. All ordered pairs of the form $(x,\ 2x-4)$

29. It looks as though the variable z would be the easiest to eliminate, so we first add the first and second equations: $(5x+y-z=17)+(3x+y+z=11)$ $\rightarrow(8x+2y=28)$. Call this Equation 1' (read "one prime"). Next we multiply the first equation by 2 and add the result to the second equation: $(10x+2y-2z=34)+(2x+5y+2z=0)\rightarrow(12x+7y=34)$. Call this Equation 2'. We now eliminate the variable x from Eqns. 1' and 2' by multiplying Eqn. 1' by 3 and Eqn. 2' by -2 (to get the LCM of 8 and 12, which is 24) and adding them: $(24x+6y=84)+(-24x-14y=-68)\rightarrow(-8y=16)$, hence $y=-2$. Substituting this into Eqn. 1' (because it is easier than substituting into Eqn. 2') we get

$$\begin{aligned} 8x+2(-2)&=28\\ 8x-4&=28\\ 8x&=32\\ x&=4. \end{aligned}$$

Now we substitute this value and the value for y into the third original equation to solve for z:

$$\begin{aligned} 3(4)+(-2)+z&=11\\ 12-2+z&=11\\ 10+z&=11\\ z&=1. \end{aligned}$$

Hence the solution is the ordered triple $(4,\ -2,\ 1)$.

33.

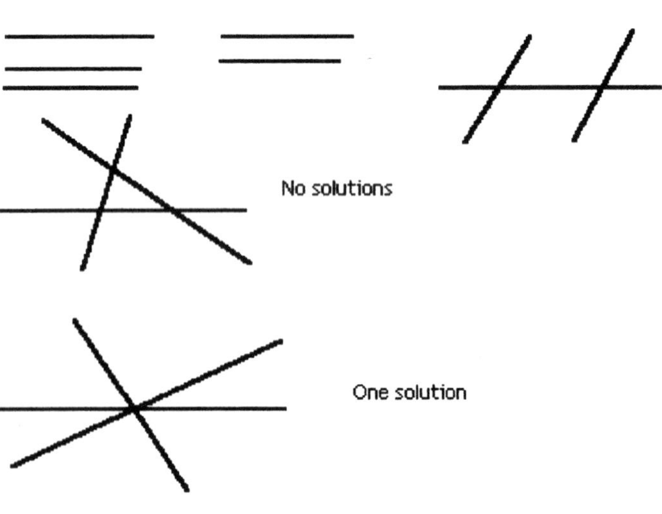

If the three equations were unique, there could be no solutions or one solution.

37. $(x, y) \approx (0.2210, 0.5034)$

Section 2.1
Introduction to Matrices and the Gauss-Jordan Method

1a. Since A has 3 rows (horizontal) and 2 columns (vertical), its dimensions are 3×2.

1b. A row (column) matrix consists of only one row (column), and a square matrix has an equal number of rows and columns, so A is none of them.

5a. Since E has 1 row and 2 columns, its dimensions are 1×2.

5b. It has one row, so it is a row matrix.

9a. Since J has 3 rows and 1 column, its dimensions are 3×1.

9b. It has one column, so it is a column matrix.

13.

$$\begin{bmatrix} c_{11} \\ c_{21} \end{bmatrix}$$

$$c_{21} = 41$$

17.

$$\begin{bmatrix} g_{11} & g_{12} & g_{13} \\ g_{21} & g_{22} & g_{23} \\ g_{31} & g_{32} & g_{33} \end{bmatrix}$$

$$g_{12} = -11$$

21.

$$\begin{bmatrix} 2 & 7 & 11 \\ 3 & -2 & 15 \end{bmatrix}$$

25.

$$\begin{bmatrix} 2 & 3 & -7 & 53 \\ 5 & -2 & 12 & 19 \\ 1 & 1 & 1 & 55 \end{bmatrix}$$

29.

$$\frac{1}{2}R_1 : R_1 \begin{bmatrix} 1 & 3 & 5 \\ -2 & 1 & 4 \end{bmatrix}$$

$$2R_1 + R_2 : R_2 \begin{bmatrix} 1 & 3 & 5 \\ 0 & 7 & 14 \end{bmatrix}$$

33.

$$\frac{1}{2}R_1 : R_1 \begin{bmatrix} 1 & 1 & 2 & 6 \\ 2 & -1 & 4 & 3 \\ 1 & 2 & -9 & 2 \end{bmatrix}$$

$$-2R_1 + R_2 : R_2 \begin{bmatrix} 1 & 1 & 2 & 6 \\ 0 & -3 & 0 & -9 \\ 1 & 2 & -9 & 2 \end{bmatrix}$$

$$-1R_1 + R_3 : R_3 \begin{bmatrix} 1 & 1 & 2 & 6 \\ 0 & -3 & 0 & -9 \\ 0 & 1 & -11 & -4 \end{bmatrix}$$

37. (0.5281, −0.6205, 0)

41. (2, 2)

45. (4, −2, 1)

49. (5, 10, −2)

53. (5.6, 9.2, 5.8)

57. (3.3333, 113.6667, 47)

61. (1.8785, 1.1574, −10.5981, 4.6650)

Section 2.2
More on the Gauss-Jordan Method

1. First we identify the variables. The question is, "How many kayaks?" Since there are two types, then there are two variables; these represent the number of kayaks of each type to be produced. Let r represent the number of river kayaks to be produced, and s (for "sea") represent the number of ocean kayaks to be produced. Now we write equations with the numerical data given; these data refer to the time available in the cutting and assembly department, and the time in each department each type of kayak requires. The cutting department is allotted 460 hours per week, and river kayaks take 40 minutes to cut, while ocean kayaks require 50 minutes to cut. Before writing equations, we must take care to convert all data into the same unit of measurement. We can convert hours to minutes, or vice-versa. Let's convert hours to minutes--460 hours $=$ 27,600 minutes. Our first equation which shows the data for the cutting department is

$$40r + 50s = 27,600.$$

Similarly, for the assembly department we have the equation

$$20r + 30s = 14,400.$$

We divide both sides of each equation by 10 to simplify, and we obtain the equivalent system of equations:

$$\begin{aligned} 4r + 5s &= 2760, \\ 2r + 3s &= 1440. \end{aligned}$$

This gives us the following matrix for solution by the Gauss-Jordan method:

$$\begin{bmatrix} 4 & 5 & 2760 \\ 2 & 3 & 1440 \end{bmatrix}.$$

$$\begin{bmatrix} 4 & 5 & 2760 \\ 2 & 3 & 1440 \end{bmatrix} \xrightarrow{\frac{1}{4}R_1:\,R_1} \begin{bmatrix} 1 & \frac{5}{4} & 690 \\ 2 & 3 & 1440 \end{bmatrix} \xrightarrow{-2R_1 + R_2:\,R_2} \begin{bmatrix} 1 & \frac{5}{4} & 690 \\ 0 & \frac{1}{2} & 60 \end{bmatrix} \xrightarrow{2R_2:\,R_2}$$

$$\begin{bmatrix} 1 & \frac{5}{4} & 690 \\ 0 & 1 & 120 \end{bmatrix} \xrightarrow{-\frac{5}{4}R_2 + R_1:\,R_1} \begin{bmatrix} 1 & 0 & 540 \\ 0 & 1 & 120 \end{bmatrix}$$

The solution is $(r,\ s) = (540,\ 120)$.

5. The variables are the number of barrels of each fertilizer type; in particular, let G represent the number of barrels of Green Beauty, N represent the number of barrels of NoMoorMan, and P represent the number of barrels of Purismal. Now we set up equations for amounts of nitrogen, phosphorus, and potassium. For nitrogen, we have $50G + 30N + 0P = 900$, and dividing both sides by 10 yields the equivalent equation $5G + 3N + 0P = 90$. For phosphorus, we have $30G + 30N + 50P = 800$, and dividing both sides by 10 yields the equivalent equation $3G + 3N + 5P = 80$. For potassium, we have $20G + 30N + 50P = 700$, and dividing both sides by 10 yields the equivalent equation $2G + 3N + 5P = 70$. Now we have a system of three equations in three unknowns, which we solve using the Gauss-Jordan method:

$$\begin{bmatrix} 5 & 3 & 0 & 90 \\ 3 & 3 & 5 & 80 \\ 2 & 3 & 5 & 70 \end{bmatrix} \xrightarrow{\frac{1}{5}R_1:\, R_1} \begin{bmatrix} 1 & \frac{3}{5} & 0 & 18 \\ 3 & 3 & 5 & 80 \\ 2 & 3 & 5 & 70 \end{bmatrix} \xrightarrow{-3R_1 + R_2:\, R_2}$$

$$\begin{bmatrix} 1 & \frac{3}{5} & 0 & 18 \\ 0 & \frac{6}{5} & 5 & 26 \\ 2 & 3 & 5 & 70 \end{bmatrix} \xrightarrow{-2R_1 + R_3:\, R_3} \begin{bmatrix} 1 & \frac{3}{5} & 0 & 18 \\ 0 & \frac{6}{5} & 5 & 26 \\ 0 & \frac{9}{5} & 5 & 34 \end{bmatrix} \xrightarrow{\frac{5}{6}R_2:\, R_2}$$

$$\begin{bmatrix} 1 & \frac{3}{5} & 0 & 18 \\ 0 & 1 & \frac{25}{6} & \frac{65}{3} \\ 0 & \frac{9}{5} & 5 & 34 \end{bmatrix} \xrightarrow{-\frac{3}{5}R_2 + R_1:\, R_1} \begin{bmatrix} 1 & 0 & -\frac{5}{2} & 5 \\ 0 & 1 & \frac{25}{6} & \frac{65}{3} \\ 0 & \frac{9}{5} & 5 & 34 \end{bmatrix} \xrightarrow{-\frac{9}{5}R_2 + R_3:\, R_3}$$

$$\begin{bmatrix} 1 & 0 & -\frac{5}{2} & 5 \\ 0 & 1 & \frac{25}{6} & \frac{65}{3} \\ 0 & 0 & -\frac{5}{2} & -5 \end{bmatrix} \xrightarrow{-\frac{2}{5}R_3:\, R_3} \begin{bmatrix} 1 & 0 & -\frac{5}{2} & 5 \\ 0 & 1 & \frac{25}{6} & \frac{65}{3} \\ 0 & 0 & 1 & 2 \end{bmatrix} \xrightarrow{-\frac{25}{6}R_3 + R_2:\, R_2}$$

$$\begin{bmatrix} 1 & 0 & -\frac{5}{2} & 5 \\ 0 & 1 & 0 & \frac{40}{3} \\ 0 & 0 & 1 & 2 \end{bmatrix} \xrightarrow{\frac{5}{2}R_3 + R_1:\, R_1} \begin{bmatrix} 1 & 0 & 0 & 10 \\ 0 & 1 & 0 & \frac{40}{3} \\ 0 & 0 & 1 & 2 \end{bmatrix}$$

The solution is $(G,\ N,\ P) = \left(10,\ \frac{40}{3},\ 2\right)$, or 10 barrels of Green Beauty, $13\frac{1}{3}$ barrels of NoMoorMan, and 2 barrels of Purismal.

9. The matrix corresponds to the system of equations:

$$x + 0y + 5z = -3,$$
$$0x + y - 7z = 8,$$
$$0x + 0y + 0z = 0.$$

The first equation gives $x = -5z - 3$ and the second gives $y = 7z + 8$. The third is true for all ordered triples (x, y, z), where x, y, z are real numbers. The solution is all ordered triples of the form $(-5z - 3,\ 7z + 8,\ z)$. We get particular solutions by setting a value for z and then computing the other two values; for example:

Let $z = 0$.
Then $x = -5(0) - 3 = -3$,
$y = 7(0) + 8 = 8$.
Solution: $(-3,\ 8,\ 0)$

13.

$$\begin{bmatrix} 8 & 2 & -1 & 21 \\ 3 & 2 & 4 & 10 \\ -7 & 2 & 14 & 20 \end{bmatrix} \to \begin{bmatrix} 1 & \frac{1}{4} & -\frac{1}{8} & \frac{21}{8} \\ 3 & 2 & 4 & 10 \\ -7 & 2 & 14 & 20 \end{bmatrix} \to \begin{bmatrix} 1 & \frac{1}{4} & -\frac{1}{8} & \frac{21}{8} \\ 0 & \frac{5}{4} & \frac{35}{8} & \frac{17}{8} \\ -7 & 2 & 14 & 20 \end{bmatrix} \to \begin{bmatrix} 1 & \frac{1}{4} & -\frac{1}{8} & \frac{21}{8} \\ 0 & \frac{5}{4} & \frac{35}{8} & \frac{17}{8} \\ 0 & \frac{15}{4} & \frac{105}{8} & \frac{307}{8} \end{bmatrix} \to \begin{bmatrix} 1 & \frac{1}{4} & -\frac{1}{8} & \frac{21}{8} \\ 0 & 1 & \frac{7}{2} & \frac{17}{10} \\ 0 & \frac{15}{4} & \frac{105}{8} & \frac{307}{8} \end{bmatrix}$$

$$\to \begin{bmatrix} 1 & 0 & -1 & \frac{11}{5} \\ 0 & 1 & \frac{7}{2} & \frac{17}{10} \\ 0 & \frac{15}{4} & \frac{105}{8} & \frac{307}{8} \end{bmatrix} \to \begin{bmatrix} 1 & 0 & -1 & \frac{11}{5} \\ 0 & 1 & \frac{7}{2} & \frac{17}{10} \\ 0 & 0 & 0 & 32 \end{bmatrix}$$

Since the last row of the final matrix corresponds to the false equation $0x + 0y + 0z = 32$, there are no solutions.

17.

$$\begin{bmatrix} 3 & -5 & 1 & 12 \\ 2 & 1 & 1 & 3 \\ 5 & -4 & 1 & 0 \end{bmatrix} \to \begin{bmatrix} 1 & 0 & 0 & -\frac{63}{13} \\ 0 & 1 & 0 & -\frac{30}{13} \\ 0 & 0 & 1 & 15 \end{bmatrix}$$

This corresponds to the solution $(x, y, z) = \left(-\frac{63}{13}, -\frac{30}{13}, 15\right)$. Substituting this into the fourth equation $x + y + z = 4$ yields

$$\begin{aligned} -\frac{63}{13} + \left(-\frac{30}{13}\right) + 15 &= \frac{102}{13} \\ &\neq 4, \end{aligned}$$

thus there is no solution.

1.

$$3x + y < 4$$
$$y < -3x + 4$$

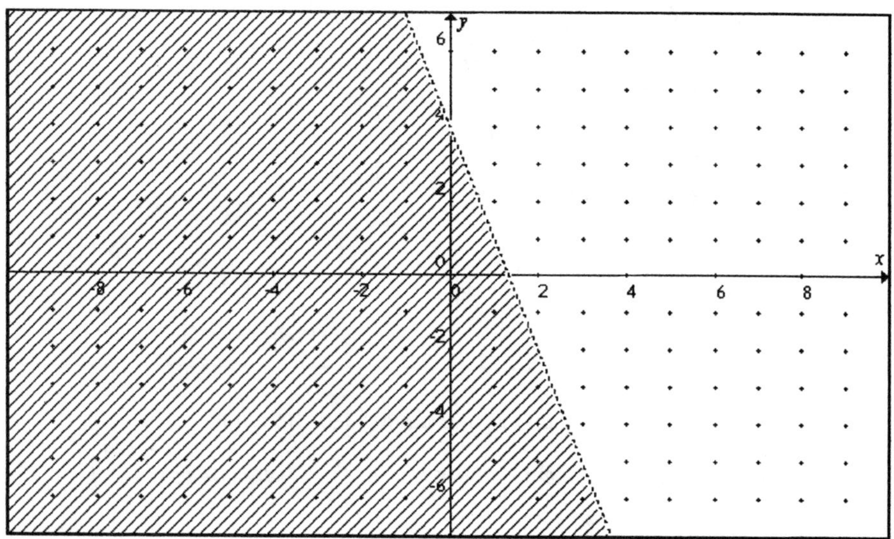

5.

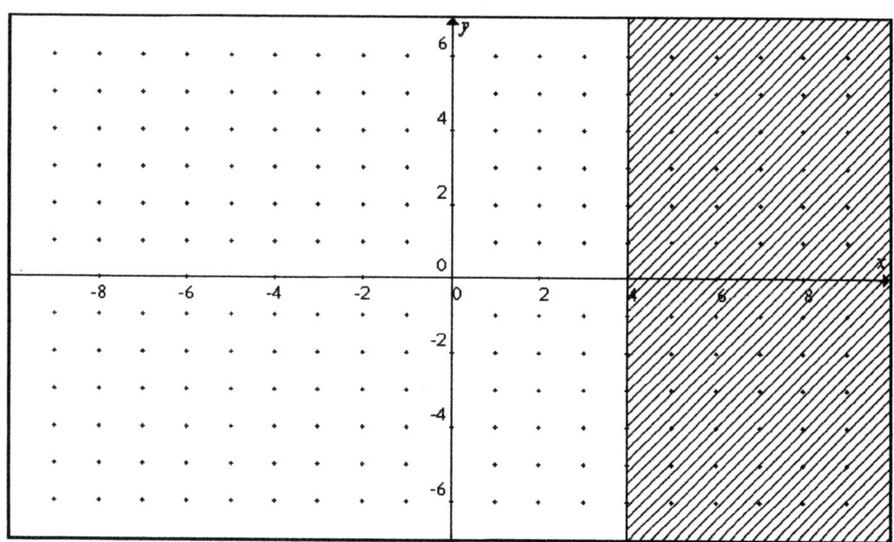

9a.

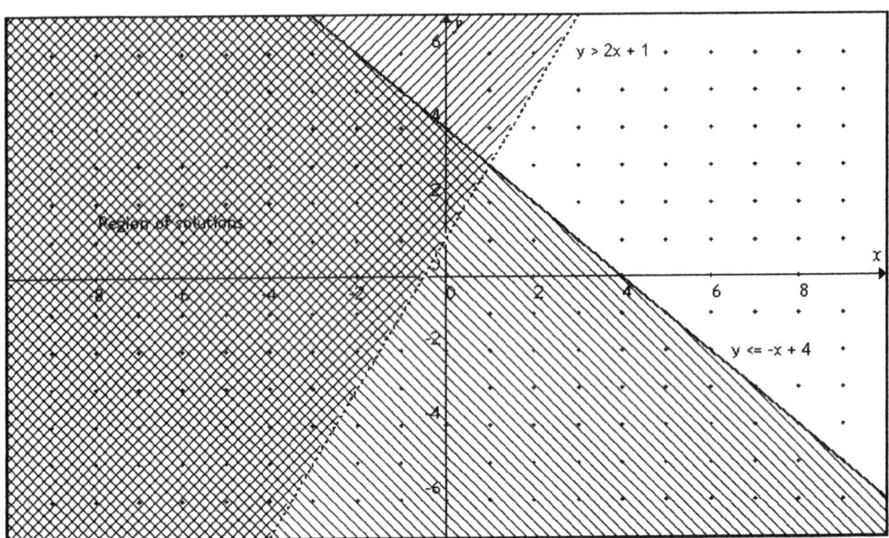

9b. The region of solutions is unbounded (to the left).

9c. The corner point occurs where $y = 2x + 1$ intersects $y = -x + 4$:

$$\begin{aligned} 2x + 1 &= -x + 4 \\ 2x + x &= 4 - 1 \\ 3x &= 3 \\ x &= 1 \\ y &= 2x + 1 \\ &= 2(1) + 1 \\ &= 3 \end{aligned}$$

The corner point is $(1, 3)$.

13a.

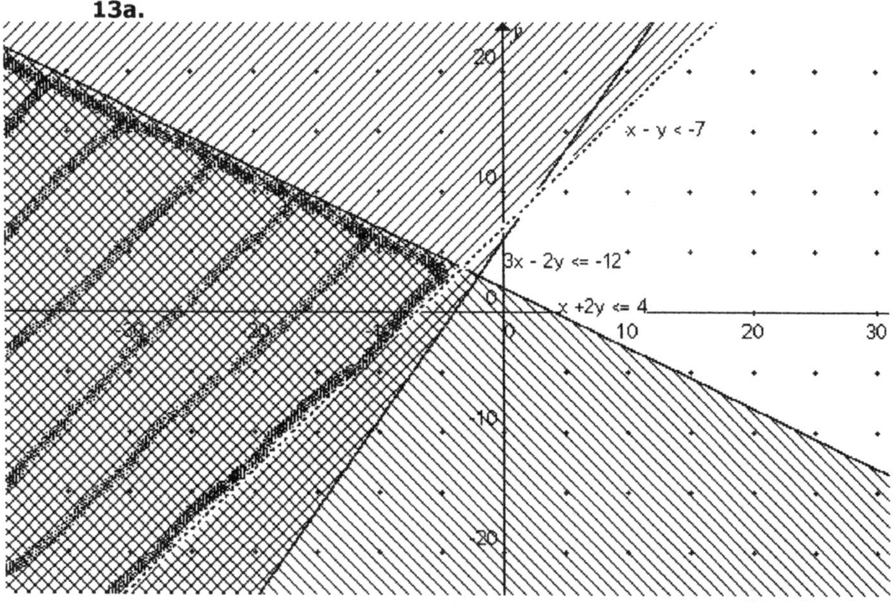

13b. The region of solutions is unbounded.

13c. It is rather difficult to see the region of solutions (and hence the corner points) at first glance. To focus in on it, first look at the region of solutions for the first two inequalities, and then note where the last inequality's solutions overlap that. That is the region of solutions for the whole system. The sole corner point occurs at the intersection of the lines $x - y = -7$ and $x + 2y = 4$. Using the Gauss-Jordan method on the matrix $\begin{bmatrix} 1 & -1 & -7 \\ 1 & 2 & 4 \end{bmatrix}$ yields $\begin{bmatrix} 1 & 0 & -\frac{10}{3} \\ 0 & 1 & \frac{11}{3} \end{bmatrix}$, which means that the corner point is $\left(-\frac{10}{3}, \frac{11}{3}\right)$.

17a.

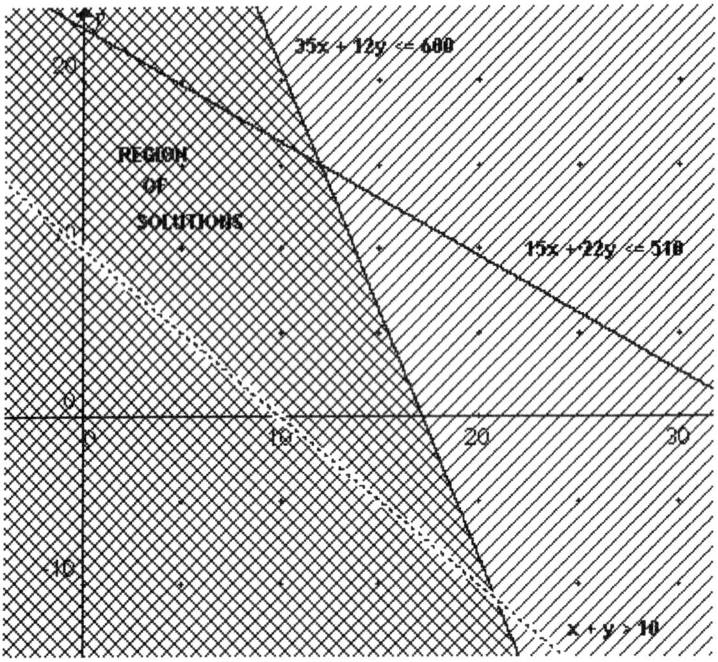

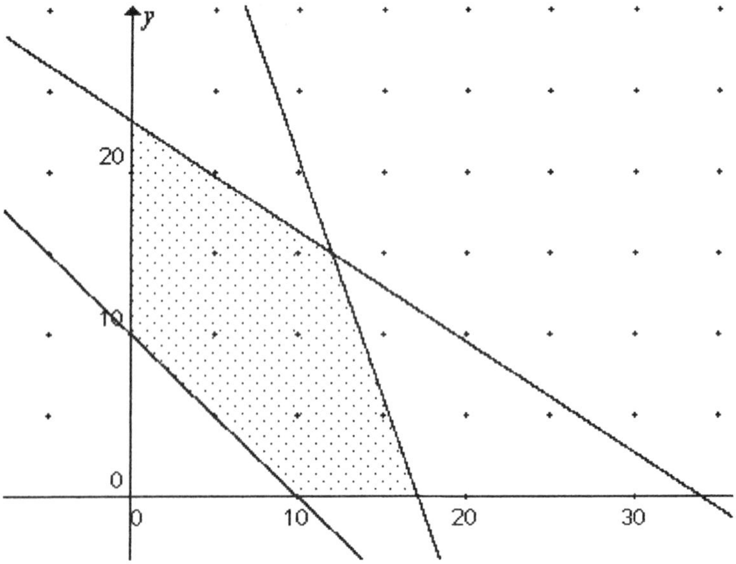

17b. The region of solutions is bounded.

17c. Note that there are 5 corner points. One occurs at the intersection of the lines $15x + 22y = 510$ and $35x + 12y = 600$. We use the Gauss-Jordan method on the matrix $\begin{bmatrix} 15 & 22 & 510 \\ 35 & 12 & 600 \end{bmatrix}$ and get $(12, 15)$ as one corner point.

Another corner point occurs where $15x + 22y = 510$ crosses the y-axis, i.e., where $x = 0$:

$$\begin{aligned} 15(0) + 22y &= 510 \\ 22y &= 510 \\ y &= \frac{510}{22} \approx 23.18 \end{aligned}$$

Thus another corner point is (0, 23.18). A corner point occurs where $35x + 12y = 600$ crosses the x-axis, i.e., where $y = 0$:

$$\begin{aligned} &35x + 12(0) = 600 \\ &35x = 600 \\ &x = \frac{600}{35} \approx 17.14 \end{aligned}$$

We now have a corner point (17.14, 0). From inspection of the graph, it is easy to see that the remaining corner points are (0, 10) and (10, 0).

21a.

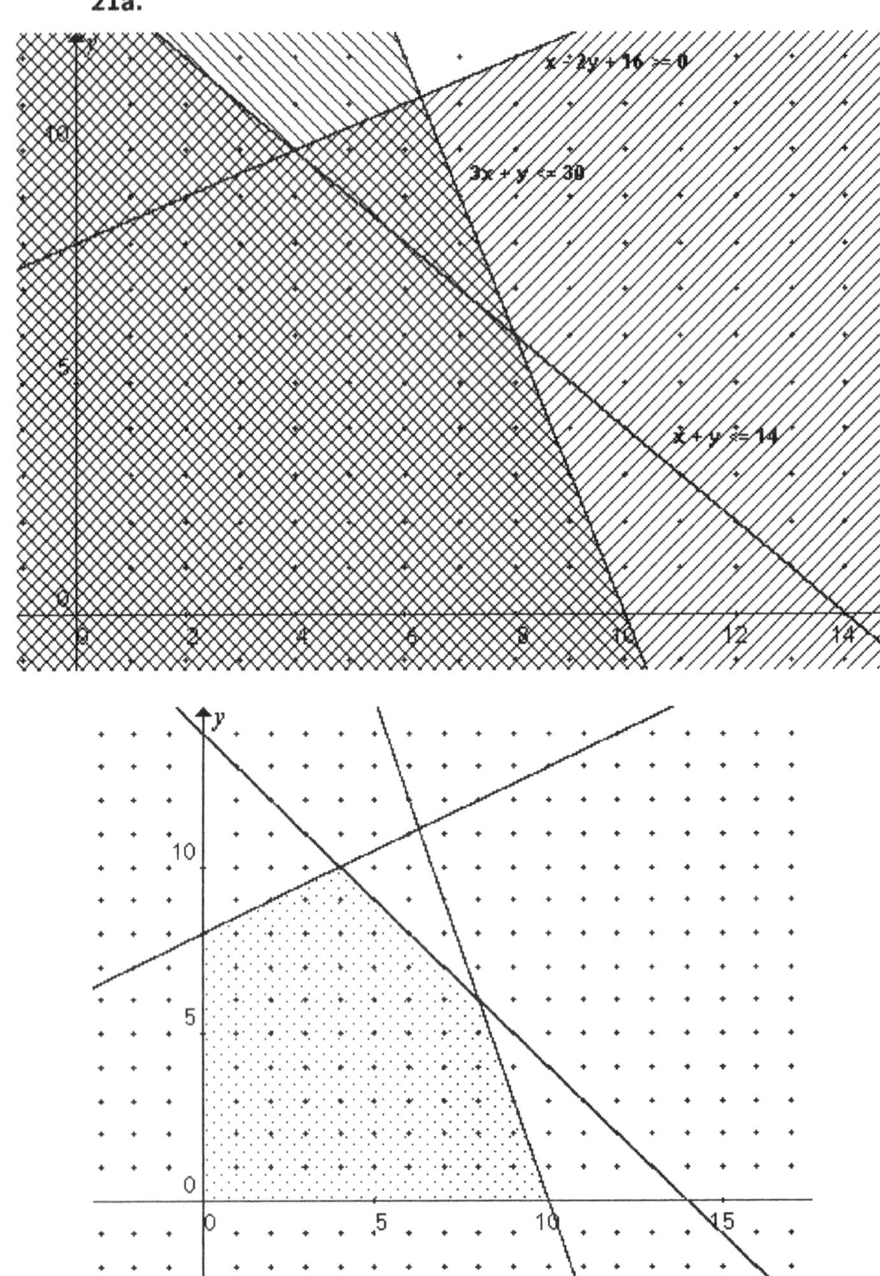

21b. The region of solutions is bounded.

21c. The corner points are easily read directly from the graph: (0, 0), (10, 0), (8, 6), (4, 10), (0, 8).

Section 2.4 The Geometry of Linear Programming

1. From the first sentence, we see that the constraint is no more than, or " $\leq$ ", 100 gallons of water per day. This means that our inequality will have " $\leq$ 100" in it. Now, each shrub and tree uses a given amount of water per day. The crucial question is, how many trees and shrubs are there? These are not given, so we use variables to represent their quantities. Thus, let s represent the number of shrubs, and t represent the number of trees. Our inequality is therefore $s + 3t \leq 100$.

5. The first sentence has the constraint; it is the amount of space available in the warehouse. Anything stored in the warehouse must take up *at most* 1650 ft^3, or " $\leq$ 1650". Let r represent the number of refrigerators to be stored in the warehouse and d represent the number of dishwashers to be stored in the warehouse. Our inequality is $63r + 41d \leq 1650$.

9. Let x represent pounds of Morning Blend coffee and y represent pounds of South American Blend coffee. First, we note that two obvious constraints are that $x \geq 0,\ y \geq 0$. Two other constraints are $\frac{1}{3}x + \frac{2}{3}y \leq 100$ (constraint on Mexican beans) and $\frac{2}{3}x + \frac{1}{3}y \leq 80$ (constraint on Columbian beans). The profit z is represented by the equation $z = 3x + 2.5y$. Now we graph the constraints:

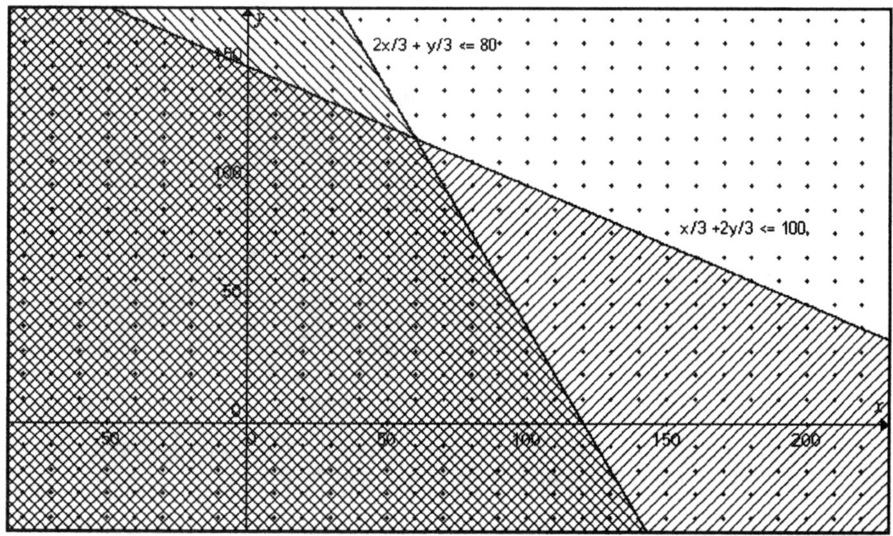

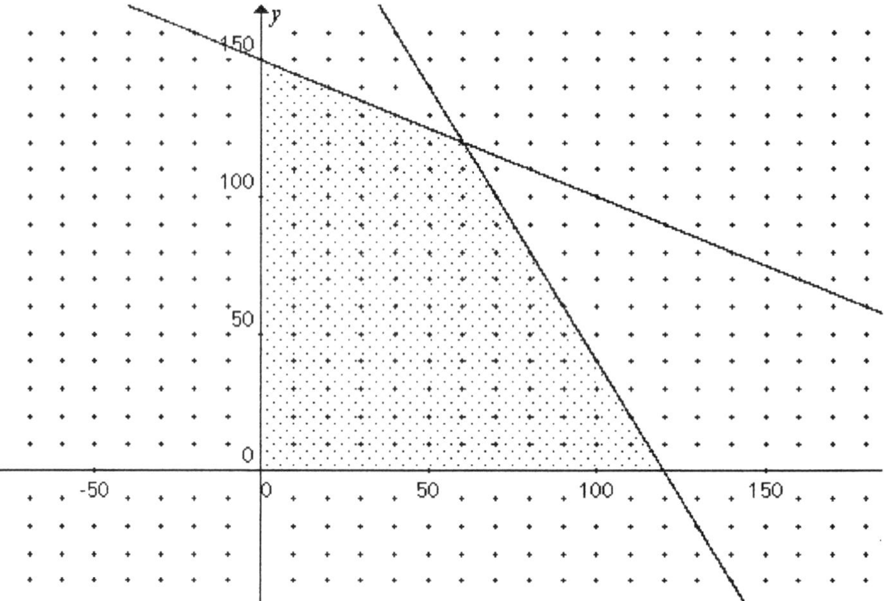

From the graph we observe that the corner points are (0, 0), (0, 150), (60, 120), and (120, 0). Also, we can find the point of intersection (60, 120) by solving the system

$$\frac{2}{3}x + \frac{1}{3}y = 80,$$
$$\frac{1}{3}x + \frac{2}{3}y = 100,$$

and we can find (120, 0) by setting $y = 0$ in the equation $\frac{2}{3}x + \frac{1}{3}y = 80$ and solving for x. (This is the x-intercept for that line.) Similarly, we can find the y-intercept of the line $\frac{1}{3}x + \frac{2}{3}y = 100$, which is (0, 150), by setting $x = 0$ and solving for y. Since the region of solutions is bounded, the maximum profit occurs at one of the corner points. We have $z = 3x + 2.5y$:

$$\text{at } (0,\ 0)\text{: } 3(0) + 2.5(0) = 0,$$
$$\text{at } (0, 150)\text{: } 3(0) + 2.5(150) = 375,$$
$$\text{at } (60,\ 120)\text{: } 3(60) + 2.5(120) = 480,$$
$$\text{at } (120,\ 0)\text{: } 3(120) + 2.5(0) = 360.$$

To realize their maximum profit of $480 they should make 60 lbs. of Morning Blend and 120 lbs. of South American Blend per day.

13. Let x represent the number of flights scheduled for the Orville 606 aircraft and y represent the number of flights scheduled for the Wilbur W-1112 aircraft. The constraints are $0 \le x \le 52,\ 0 \le y \le 30,\ 20x + 80y \ge 1600,$ and $80x + 120y \ge 4800$. The objective function, reflecting the operating costs, is $z = 12{,}000x + 18{,}000y$ and must be minimized.

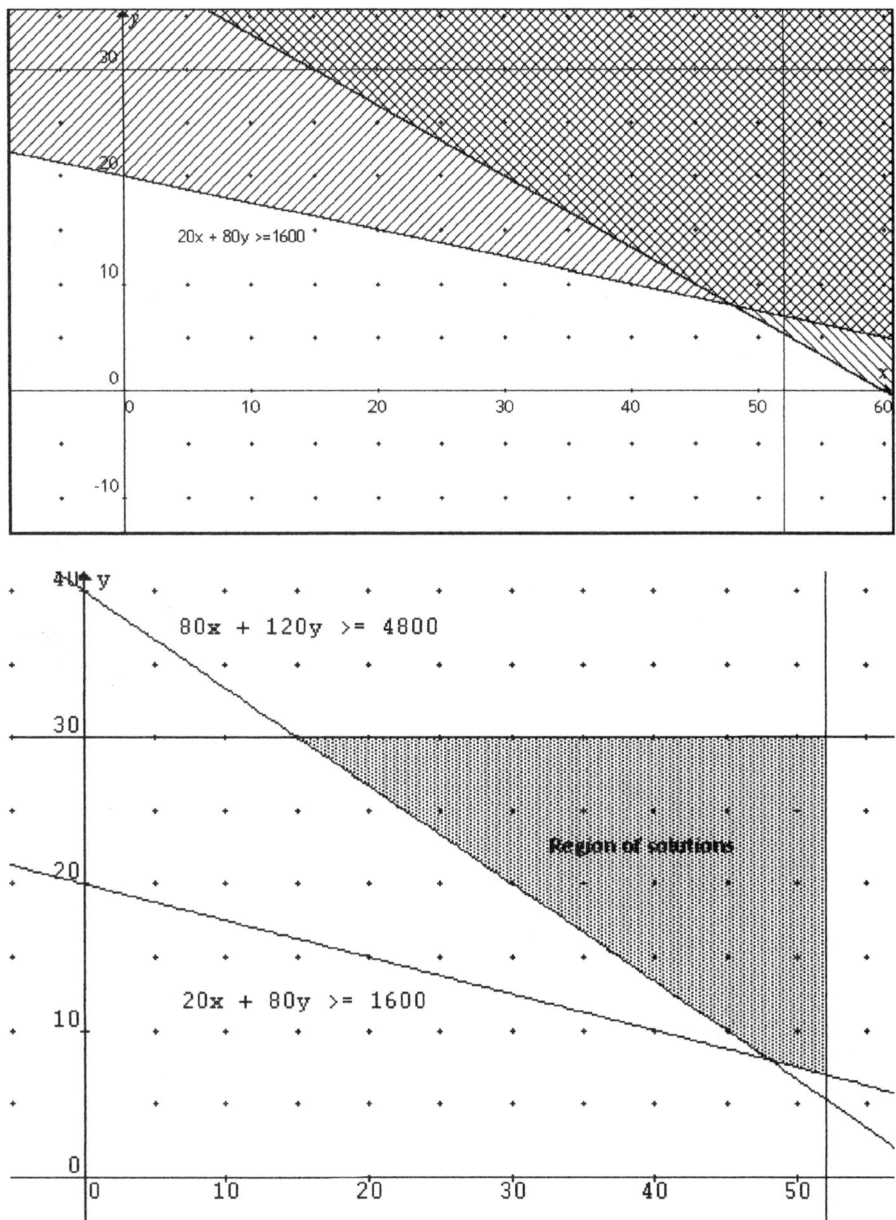

The corner points are (48, 8), (15, 30), (52, 7), and (52, 30). Minimum operating cost of 720,000 occurs at $(48, 8)$ and at $(15, 30)$, which means that the minimum cost occurs at any point on the line joining these two points, $y = -\frac{2}{3}x + 40$, with domain $15 \le x \le 48$.

17. To maximize profit, recall that 60 pounds of Morning Blend and 120 pounds of South American Blend should be made. The shop obtains 100 pounds of Mexican beans and 80 pounds of Colombian beans daily. Maximum-profit

Morning Blend consists of $\frac{1}{3}(60) = 20$ lbs. Mexican beans and 40 lbs. Colombian beans, and maximum-profit South American blend consists of $\frac{2}{3}(120) = 80$ lbs. Mexican beans and 40 lbs. Colombian beans, for a total of 100 lbs. of Mexican beans and 80 lbs. of Colombian beans. Thus all of the beans are used.

29.

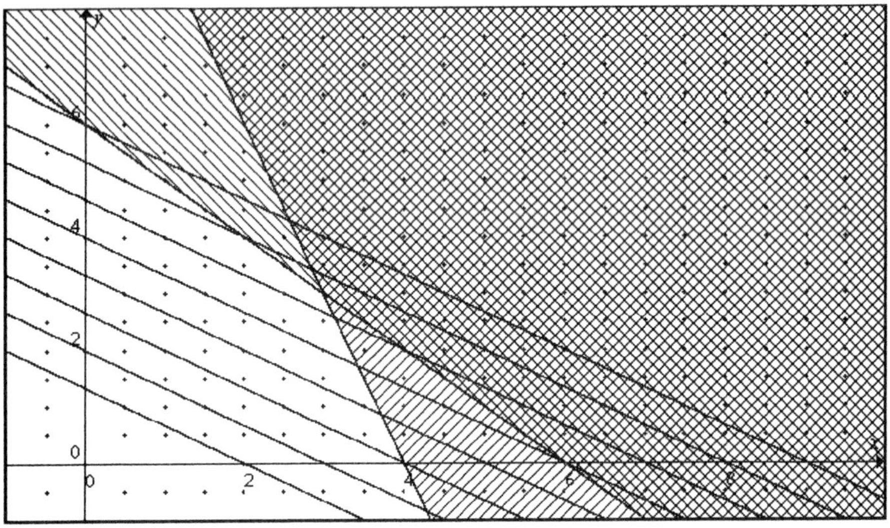

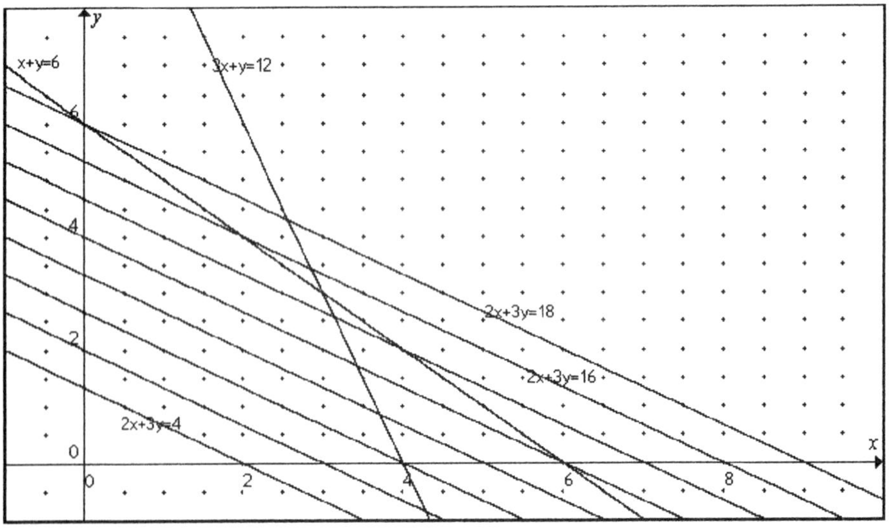

The parallel lines are, from left to right, 2x+3y=4, 2x+3y=6, 2x+3y=8, ..., 2x+3y=16, 2x+3y=18

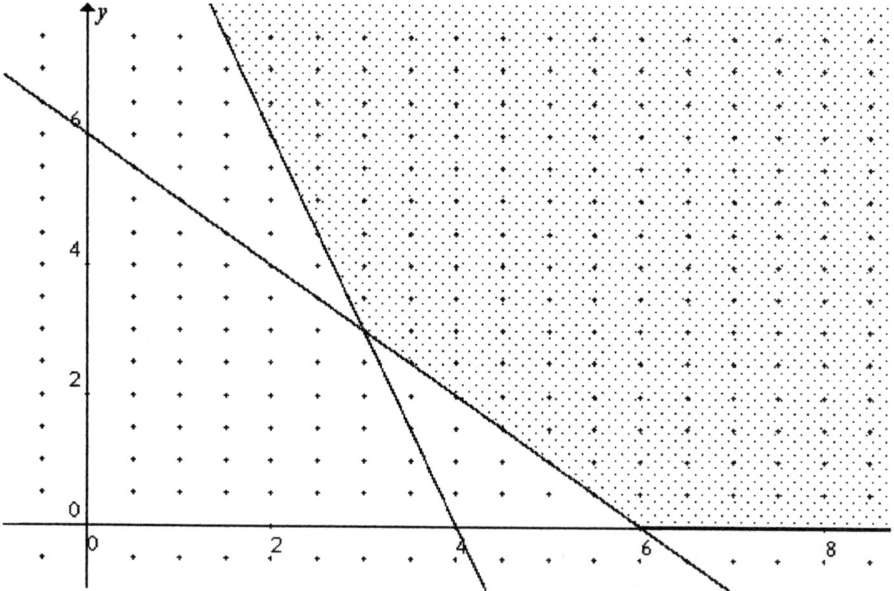

Region of solutions

The corner points occur at (3, 3), (0, 12), and (6, 0). The point at which the minimum occurs is (6, 0). There is no maximum.

Chapter 2
Review Exercises

1. Note that the second equation is three times the third equation. There is one solution; the two lines intersect at a point.

2. No solution. Observe that the left-hand side of the first equation is four times the left-hand side of the third equation, but no such relation holds on the right-hand sides of these equations--thus we have a contradiction. These lines are parallel and do not intersect.

3. Since the second equation is two times the first equation, they are equivalent, and there is essentially one line. There are an infinite number of solutions, all of them being all of the points on this line.

4. (3, −1)
Hint: Multiply the first equation by -2 and add the two equations to eliminate the variable y; then solve for x and finally solve for y.

5. All ordered pairs of the form $(x,\ -2x+4)$; (0, 4), (1, 2), (−1, 6)
Hint: Eliminate the y variable by noting that the least common multiple (LCM) of 4 and 3 is 12; multiply the first equation by 3 and the second equation by -4.

6. No solutions.
Hint: Divide the first equation by 3 and the second equation by 2 to see why.

7. Multiply the second equation by 2 and add to the first equation to eliminate y:

$$2(3x+y-4z)=2(9)$$
$$6x+2y-8z=18$$

$$(6x+2y-8z)+(8x-2y+z)=18+19$$
$$14x-7z=37 \qquad \text{(Eqn. 1}'\text{)}$$

Now multiply the second equation by 3 and add to the third equation to eliminate y:

$$3(3x+y-4z)=3(9)$$
$$9x+3y-12z=27$$

$$(9x+3y-12z)+(5x-3y+3z)=27+13$$
$$14x-9z=40 \qquad \text{(Eqn. 2}'\text{)}$$

Now solve Equations 1′ and 2′ as usual for x and z, and substitute these values into one of the original equations to solve for y. The solution is $\left(\frac{53}{28}, -\frac{75}{28}, -\frac{3}{2}\right)$.

8. This has 3 rows and 2 columns; thus its dimensions are 3×2.

9. This is a 3×3 square matrix.

10. This has 3 rows and 1 column; thus its dimensions are 3×1 and it is a column matrix.

11. This has 1 row and 2 columns; thus its dimensions are 1×2 and it is a row matrix.

12.

$$\begin{bmatrix} 5 & -6 & 2 & 6 \\ 4 & 5 & -2 & 11 \\ 6 & 1 & 3 & 16 \end{bmatrix} \to \begin{bmatrix} 1 & -\frac{6}{5} & \frac{2}{5} & \frac{6}{5} \\ 4 & 5 & -2 & 11 \\ 6 & 1 & 3 & 16 \end{bmatrix} \to \begin{bmatrix} 1 & -\frac{6}{5} & \frac{2}{5} & \frac{6}{5} \\ 0 & \frac{49}{5} & -\frac{18}{5} & \frac{31}{5} \\ 6 & 1 & 3 & 16 \end{bmatrix} \to \begin{bmatrix} 1 & -\frac{6}{5} & \frac{2}{5} & \frac{6}{5} \\ 0 & \frac{49}{5} & -\frac{18}{5} & \frac{31}{5} \\ 0 & \frac{41}{5} & \frac{3}{5} & \frac{44}{5} \end{bmatrix} \to \begin{bmatrix} 1 & -\frac{6}{5} & \frac{2}{5} & \frac{6}{5} \\ 0 & 1 & -\frac{18}{49} & \frac{31}{49} \\ 0 & \frac{41}{5} & \frac{3}{5} & \frac{44}{5} \end{bmatrix}$$

$$\to \begin{bmatrix} 1 & 0 & -\frac{2}{49} & \frac{96}{49} \\ 0 & 1 & -\frac{18}{49} & \frac{31}{49} \\ 0 & \frac{41}{5} & \frac{3}{5} & \frac{44}{5} \end{bmatrix} \to \begin{bmatrix} 1 & 0 & -\frac{2}{49} & \frac{96}{49} \\ 0 & 1 & -\frac{18}{49} & \frac{31}{49} \\ 0 & 0 & \frac{177}{49} & \frac{177}{49} \end{bmatrix} \to \begin{bmatrix} 1 & 0 & -\frac{2}{49} & \frac{96}{49} \\ 0 & 1 & -\frac{18}{49} & \frac{31}{49} \\ 0 & 0 & 1 & 1 \end{bmatrix} \to \begin{bmatrix} 1 & 0 & -\frac{2}{49} & \frac{96}{49} \\ 0 & 1 & 0 & 1 \\ 0 & 0 & 1 & 1 \end{bmatrix} \to \begin{bmatrix} 1 & 0 & 0 & 2 \\ 0 & 1 & 0 & 1 \\ 0 & 0 & 1 & 1 \end{bmatrix}$$

Therefore $(x, y, z) = (2, 1, 1)$.

13.

$$\begin{bmatrix} 3 & -1 & 2 & 18 \\ 2 & 1 & 1 & 9 \\ 10 & 0 & 6 & 54 \end{bmatrix} \to \begin{bmatrix} 1 & -\frac{1}{3} & \frac{2}{3} & 6 \\ 2 & 1 & 1 & 9 \\ 10 & 0 & 6 & 54 \end{bmatrix} \to \begin{bmatrix} 1 & -\frac{1}{3} & \frac{2}{3} & 6 \\ 0 & \frac{5}{3} & -\frac{1}{3} & -3 \\ 10 & 0 & 6 & 54 \end{bmatrix} \to \begin{bmatrix} 1 & -\frac{1}{3} & \frac{2}{3} & 6 \\ 0 & \frac{5}{3} & -\frac{1}{3} & -3 \\ 0 & \frac{10}{3} & -\frac{2}{3} & -6 \end{bmatrix} \to \begin{bmatrix} 1 & -\frac{1}{3} & \frac{2}{3} & 6 \\ 0 & 1 & -\frac{1}{5} & -\frac{9}{5} \\ 0 & \frac{10}{3} & -\frac{2}{3} & -6 \end{bmatrix}$$

$$\to \begin{bmatrix} 1 & 0 & \frac{3}{5} & \frac{27}{5} \\ 0 & 1 & -\frac{1}{5} & -\frac{9}{5} \\ 0 & \frac{10}{3} & -\frac{2}{3} & -6 \end{bmatrix} \to \begin{bmatrix} 1 & 0 & \frac{3}{5} & \frac{27}{5} \\ 0 & 1 & -\frac{1}{5} & -\frac{9}{5} \\ 0 & 0 & 0 & 0 \end{bmatrix}$$

From the final matrix we see that

$$x + 0y + \frac{3}{5}z = \frac{27}{5}$$
$$x = -\frac{3}{5}z + \frac{27}{5},$$
$$0x + y - \frac{1}{5}z = -\frac{9}{5}$$
$$y = \frac{1}{5}z - \frac{9}{5}.$$

Thus $(x, y, z) = (-\frac{3}{5}z + \frac{27}{5}, \frac{1}{5}z - \frac{9}{5}, z)$; for a particular solution let z equal any real-number value and then solve for x and y.

14.

$$\begin{bmatrix} 8 & 2 & -4 & 2 \\ 3 & 2 & 4 & 10 \\ 14 & 6 & 4 & 12 \end{bmatrix} \to \begin{bmatrix} 1 & \frac{1}{4} & -\frac{1}{2} & \frac{1}{4} \\ 3 & 2 & 4 & 10 \\ 14 & 6 & 4 & 12 \end{bmatrix} \to \begin{bmatrix} 1 & \frac{1}{4} & -\frac{1}{2} & \frac{1}{4} \\ 0 & \frac{5}{4} & \frac{11}{2} & \frac{37}{4} \\ 14 & 6 & 4 & 12 \end{bmatrix} \to \begin{bmatrix} 1 & \frac{1}{4} & -\frac{1}{2} & \frac{1}{4} \\ 0 & \frac{5}{4} & \frac{11}{2} & \frac{37}{4} \\ 0 & \frac{5}{2} & 11 & \frac{17}{2} \end{bmatrix} \to \begin{bmatrix} 1 & \frac{1}{4} & -\frac{1}{2} & \frac{1}{4} \\ 0 & 1 & \frac{22}{5} & \frac{37}{5} \\ 0 & \frac{5}{2} & 11 & \frac{17}{2} \end{bmatrix}$$

$$\to \begin{bmatrix} 1 & 0 & -\frac{8}{5} & -\frac{8}{5} \\ 0 & 1 & \frac{22}{5} & \frac{37}{5} \\ 0 & \frac{5}{2} & 11 & \frac{17}{2} \end{bmatrix} \to \begin{bmatrix} 1 & 0 & -\frac{8}{5} & -\frac{8}{5} \\ 0 & 1 & \frac{22}{5} & \frac{37}{5} \\ 0 & 0 & 0 & -10 \end{bmatrix}$$

The last row in the final matrix represents the false equation $0x + 0y + 0z = -10$; thus there is no solution.

15. Let x represent number of DoggYums tablets and y represent number of Arf-O-Vite tablets. The system of equations for this problem is:

$$4x + 5y = 40,$$
$$5x + 3y = 37.$$

The solution is (5, 4).

16.

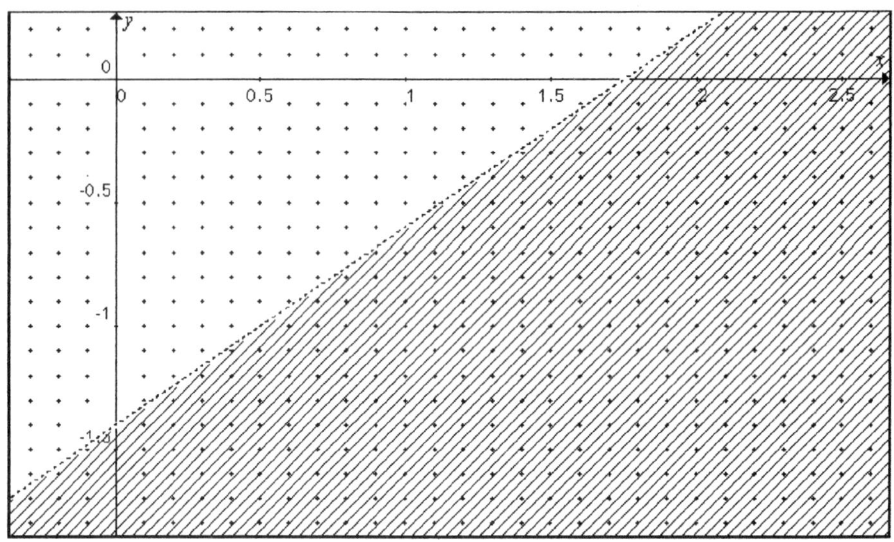

$4x - 5y > 7$, or $y < \frac{4}{5}x - \frac{7}{5}$

17.

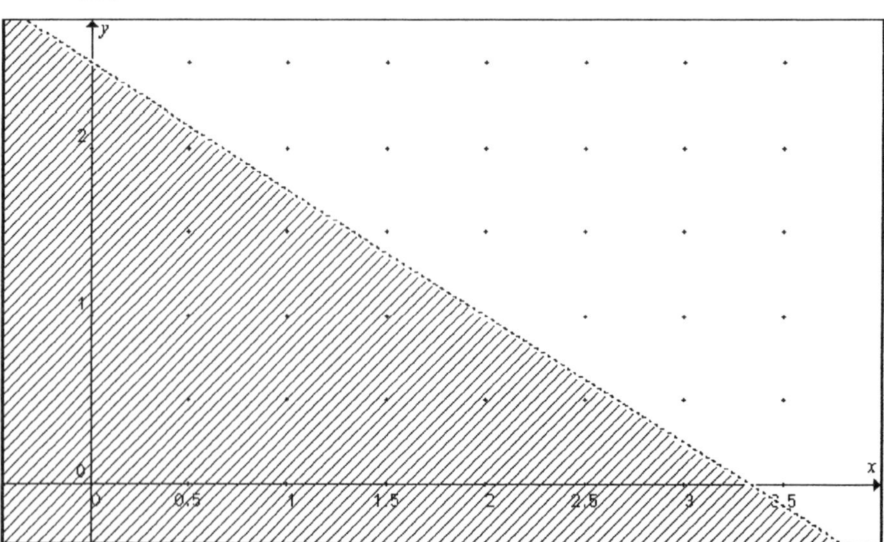

$3x + 4y < 10$, or $y < -\frac{3}{4}x + \frac{5}{2}$

18.

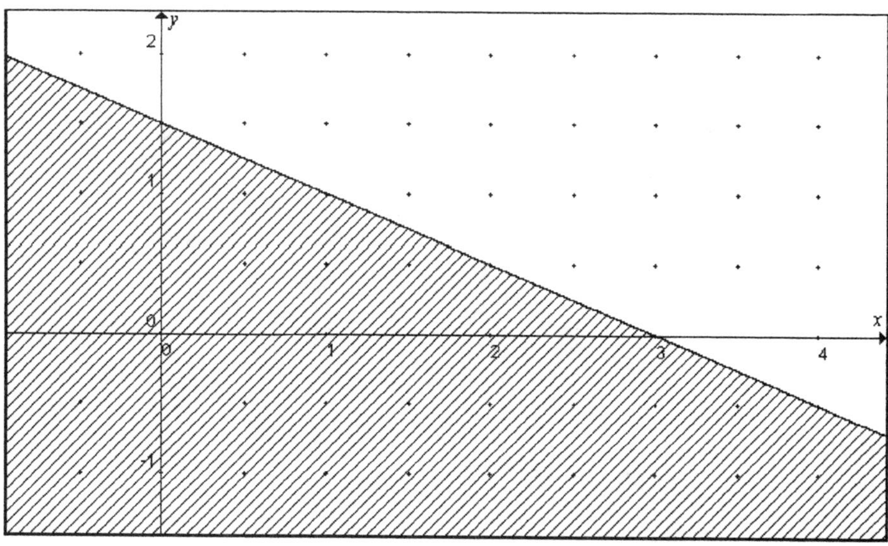

$3x + 6y \leq 9$, or $y \leq -\frac{1}{2}x + \frac{3}{2}$

19.

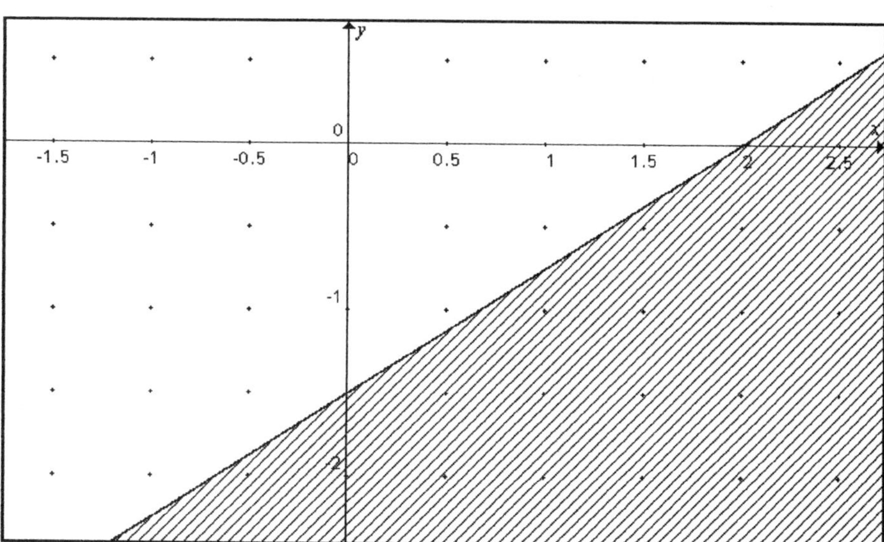

$6x - 8y \geq 12$, or $y \leq \frac{3}{4}x - \frac{3}{2}$

20.

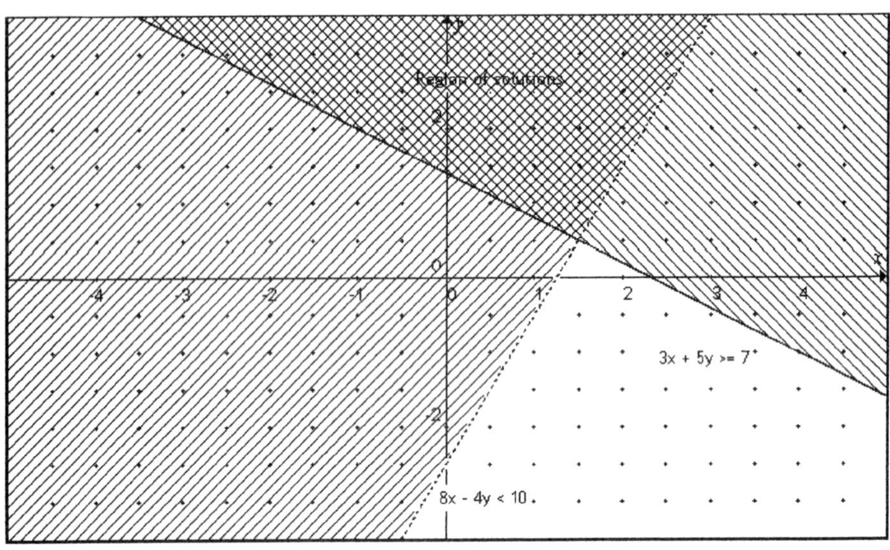

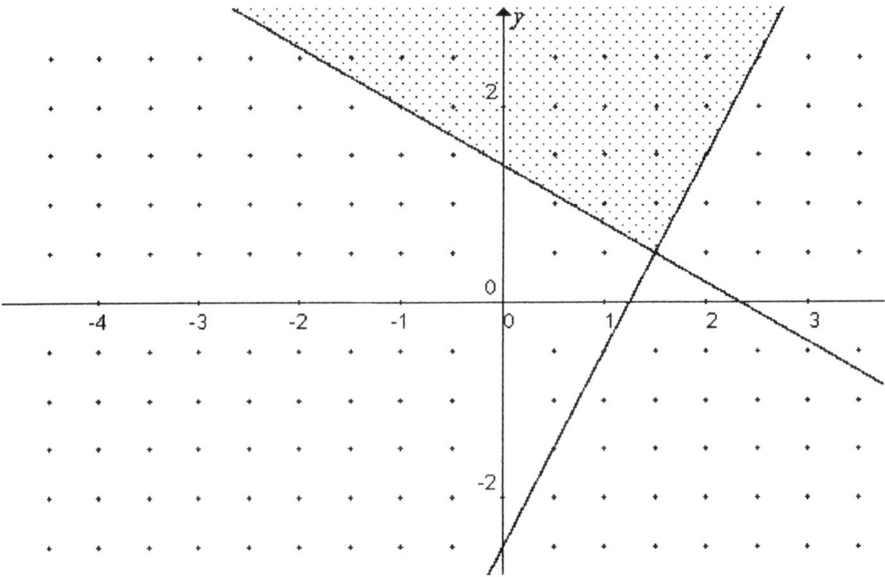

Corner point $\left(\frac{3}{2}, \frac{1}{2}\right)$.

21.

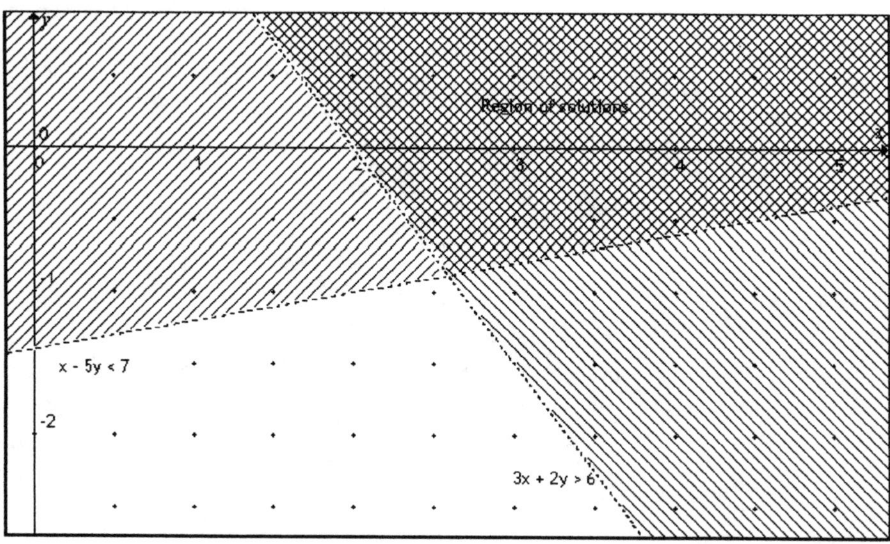

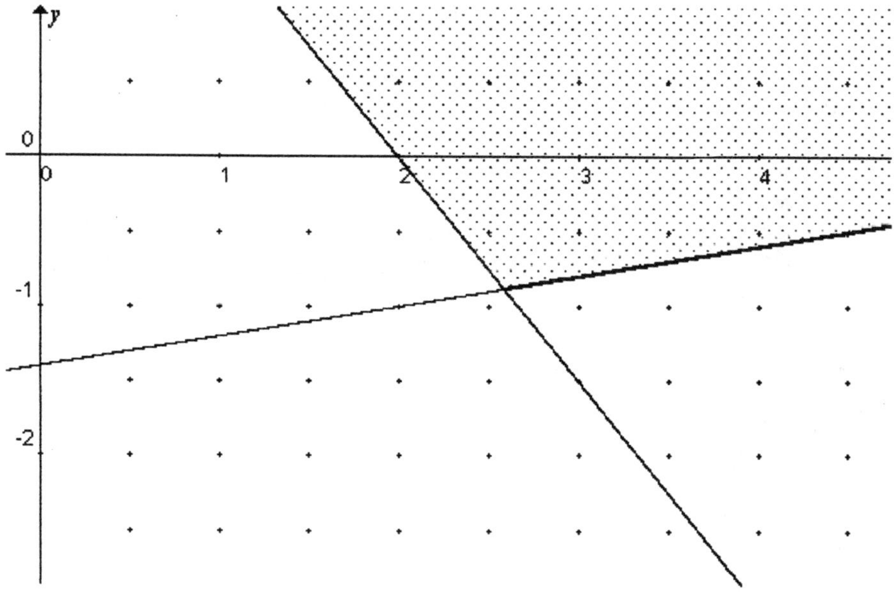

Corner point (2.59, −0.88).

22.

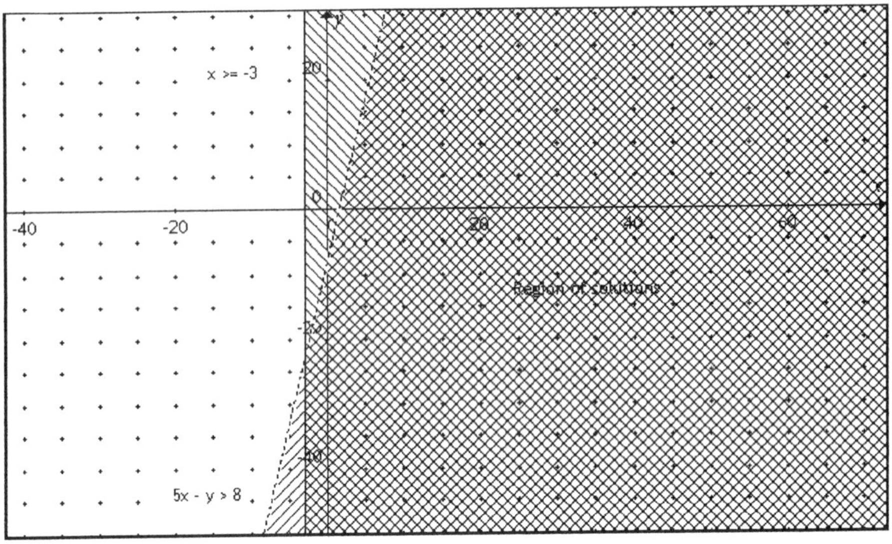

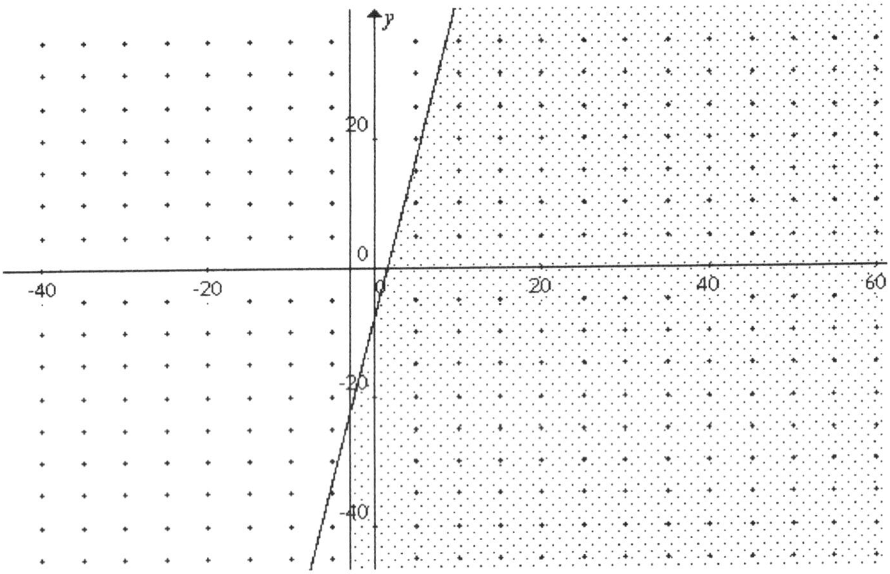

Corner point (−3, −23).

23.

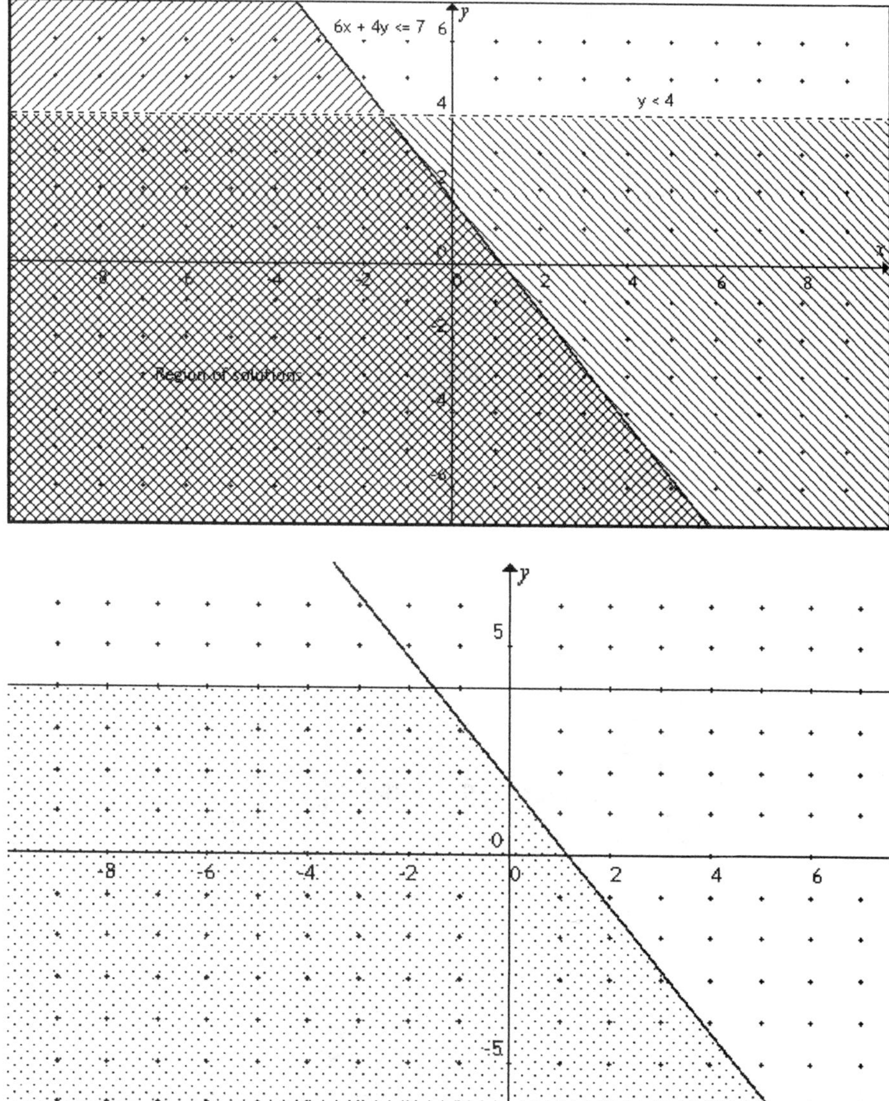

Corner point $\left(-\frac{3}{2}, 4\right)$.

24.

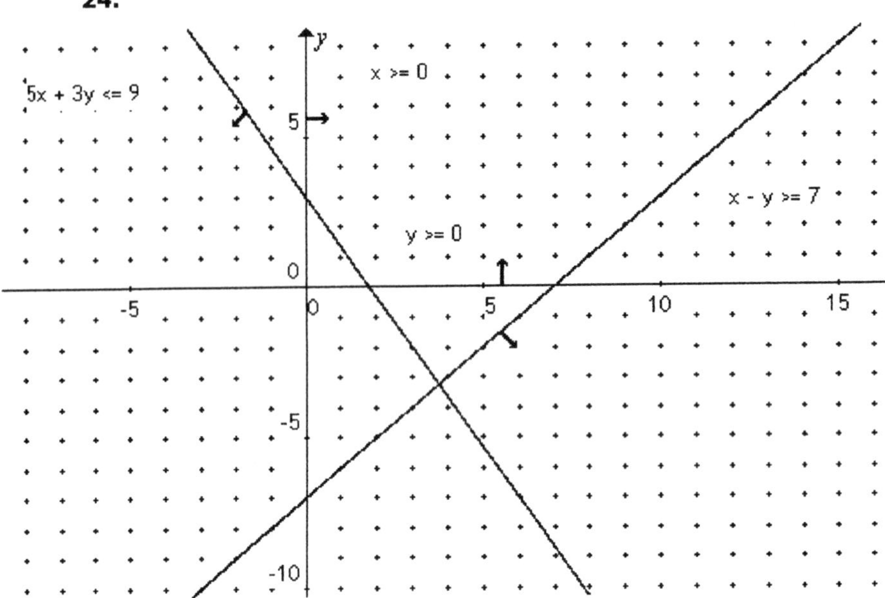

No solutions.

25.

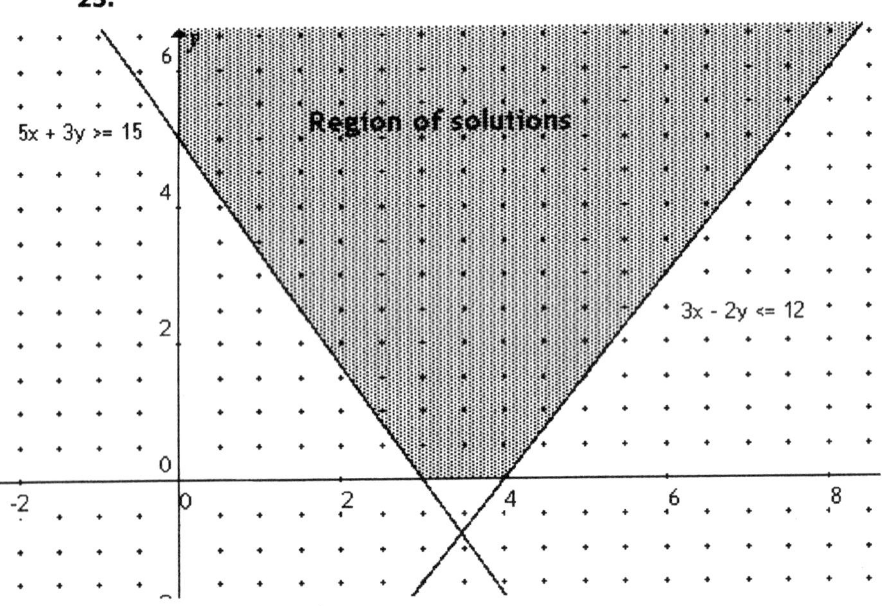

Corner points (0, 5), (3, 0), and (4, 0).

26. Let x represent the number of Model 110 speaker assemblies and y represent the number of Model 330 speaker assemblies. The constraints are:

$$\begin{aligned} x + 2y &\leq 90, \\ x + y &\leq 60, \\ 0 \leq y &\leq 44, \\ x &\geq 0. \end{aligned}$$

The objective function has to do with income, and it is to be maximized:

$$z = 200x + 350y.$$

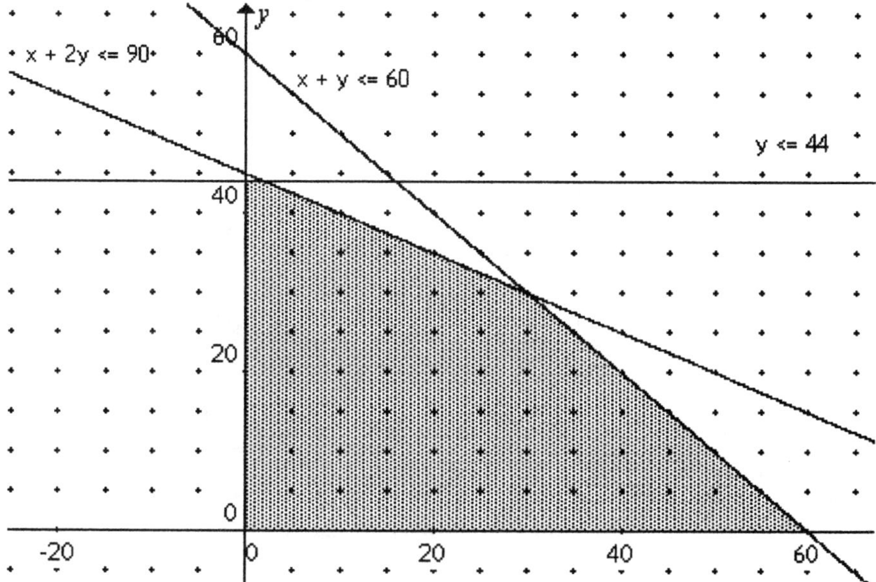

The five corner points are (0, 0), (60, 0), (30, 30), (2, 44), and (0, 44). The maximum income of $16,500 occurs at (30, 30).

27. Let x represent the number of Sunnys stocked each month and y represent the number of Iwas stocked each month. We then have the following constraints:

$$\begin{aligned} x &\geq 0, \\ y &\geq 0, \\ x + y &\geq 600, \\ x &\geq 2y, \\ 12x + 8y &\leq 12,000. \end{aligned}$$

We have an objective function for profit:

$$z = 220x + 200y.$$

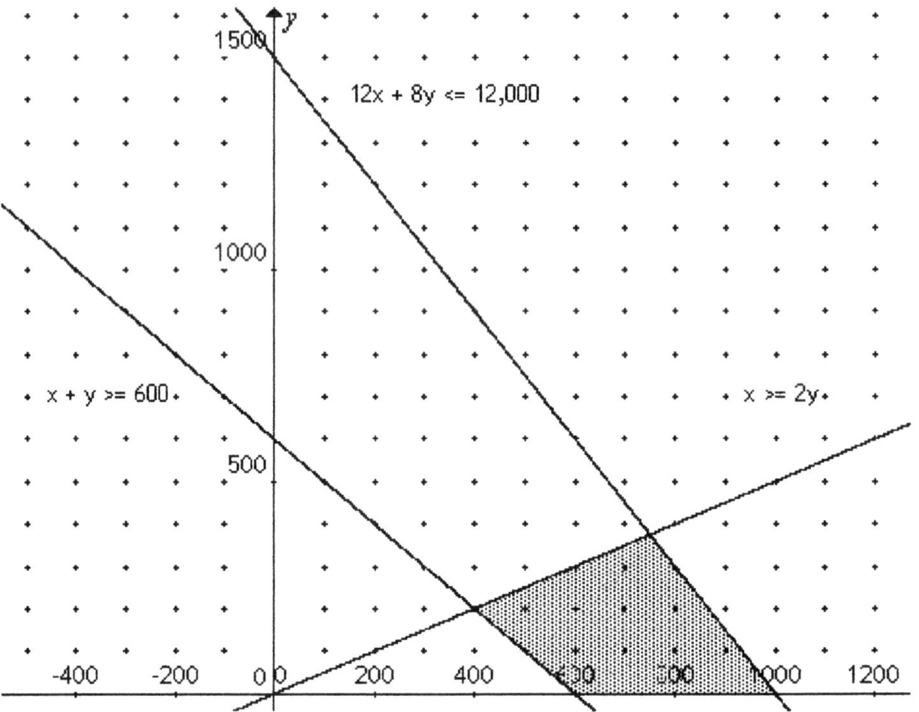

The four corner points are (600, 0), (400, 200), (750, 375), and (1000, 0). The maximum profit of $240,000 occurs at (750, 375).

Chapter 3 Linear Programming: The Simplex Method

Section 3.1 Introduction to the Simplex Method

1. $3x_1 + 2x_2 + s = 5$, where $s \geq 0$ is the slack variable. Note: There is no need to use a subscript with the slack variable here (as in "s_1") because there is only one slack variable in this exercise. A subscript would be needed to distinguish among more than one slack variable, such as is the case in the text's examples.

5a. $6.99x_1 + 3.15x_2 + 1.98x_3 \leq 42.15$

5b. $6.99x_1 + 3.15x_2 + 1.98x_3 + s = 42.15$

5c. Pounds of meat are represented by x_1, pounds of cheese are represented by x_2, loaves of bread are represented by x_3, and the slack variable s represents unused cash.

9a. $24x_1 + 36x_2 + 56x_3 + 72x_4 \leq 50,000$

9b. $24x_1 + 36x_2 + 56x_3 + 72x_4 + s = 50,000$

9c. Number of twin beds is represented by x_1, number of double beds is represented by x_2, number of queen-size beds is represented by x_3, and number of king-size beds is represented by x_4. The slack variable s represents unused warehouse space in square feet.

13. $(x_1, x_2, s_1, s_2, s_3) = (0, 0, 7.8, 9.3, 0.5)$, $z = 9.6$

17. We rewrite the constraints and the objective function:

$$\begin{aligned} 3x_1 + 4x_2 + s_1 + 0s_2 + 0z &= 40, \\ 4x_1 + 7x_2 + 0s_1 + s_2 + 0z &= 50, \\ -2x_1 - 4x_2 + 0s_1 + 0s_2 + z &= 0. \end{aligned}$$

This gives us the first simplex matrix

$$\begin{bmatrix} 3 & 4 & 1 & 0 & 0 & 40 \\ 4 & 7 & 0 & 1 & 0 & 50 \\ -2 & -4 & 0 & 0 & 1 & 0 \end{bmatrix}$$

with possible solutions $(x_1, x_2, s_1, s_2) = (0, 0, 40, 50)$, $z = 0$.

21. We rewrite the constraints and the objective function:

$$\begin{aligned} 5x_1 + 3x_2 + 9x_3 + s_1 + 0s_2 + 0s_3 + 0z &= 10, \\ 12x_1 + 34x_2 + 100x_3 + 0s_1 + s_2 + 0s_3 + 0z &= 10, \\ 52x_1 + 7x_2 + 12x_3 + 0s_1 + 0s_2 + s_3 + 0z &= 10, \\ -4x_1 - 7x_2 - 9x_3 + 0s_1 + 0s_2 + 0s_3 + z &= 0. \end{aligned}$$

This gives us the first simplex matrix

$$\begin{bmatrix} 5 & 3 & 9 & 1 & 0 & 0 & 0 & 10 \\ 12 & 34 & 100 & 0 & 1 & 0 & 0 & 10 \\ 52 & 7 & 12 & 0 & 0 & 1 & 0 & 10 \\ -4 & -7 & -9 & 0 & 0 & 0 & 1 & 0 \end{bmatrix}$$

with possible solutions $(x_1,\ x_2,\ x_3, s_1,\ s_2,\ s_3) = (0,\ 0,\ 0,\ 10,\ 10,\ 10)$, $z = 0$.

25. Let x_1 represent pounds of Yusip Blend and x_2 represent pounds of Exotic Blend. The pounds of Costa Rican beans in the Yusip Blend and the pounds of Costa Rican beans in the Exotic Blend cannot exceed 200 pounds; i.e., $\frac{1}{2}x_1 + \frac{1}{4}x_2 \leq 200$. The pounds of Ethiopian beans in the Yusip Blend and the pounds of Ethiopian beans in the Exotic Blend cannot exceed 330 pounds; i.e., $\frac{1}{2}x_1 + \frac{3}{4}x_2 \leq 330$. We want to maximize the profit from the Yusip Blend and from the Exotic Blend; i.e., we want to maximize $z = 3.5x_1 + 4x_2$. Rewriting the constraints and the objective function, we obtain

$$\begin{aligned} \frac{1}{2}x_1 + \frac{1}{4}x_2 + s_1 + 0s_2 + 0z &= 200, \\ \frac{1}{2}x_1 + \frac{3}{4}x_2 + 0s_1 + s_2 + 0z &= 330, \\ -3.5x_1 - 4x_2 + 0s_1 + 0s_2 + z &= 0. \end{aligned}$$

This gives us the first simplex matrix

$$\begin{bmatrix} \frac{1}{2} & \frac{1}{4} & 1 & 0 & 0 & 200 \\ \frac{1}{2} & \frac{3}{4} & 0 & 1 & 0 & 330 \\ -3.5 & -4 & 0 & 0 & 1 & 0 \end{bmatrix}$$

with possible solutions $(x_1,\ x_2,\ s_1,\ s_2) = (0,\ 0,\ 200,\ 330)$, $z = 0$.

Section 3.2
The Simplex Method: Complete Problems

1. The pivot column is the column containing the most negative entry in the objective function row; in this problem it is column 3. Dividing the last entry in each constraint row by the corresponding entry in the pivot column, we obtain that

Row 1, column 3: $\dfrac{12}{3} = 4,$

Row 2, column 3: $\dfrac{22}{19} \approx 1.2.$

Since the smallest nonnegative quotient is derived from entries in the second row, row 2 is the pivot row. Thus row 2, column 3 is the pivoting entry.

5a. The pivot column is the column containing the most negative entry in the objective function row; in this problem it is column 1. Dividing the last entry in each constraint row by the corresponding entry in the pivot column, we obtain that

Row 1, column 1: $\dfrac{3}{1} = 3,$

Row 2, column 1: $\dfrac{2}{4} = \dfrac{1}{2}.$

Since the smallest nonnegative quotient is derived from entries in the second row, row 2 is the pivot row. Thus row 2, column 1 is the pivoting entry.

5b.

$$\frac{1}{4} \cdot R_2 \begin{bmatrix} 1 & 2 & 1 & 0 & 0 & 3 \\ 1 & \frac{1}{4} & 0 & \frac{1}{4} & 0 & \frac{1}{2} \\ -6 & -4 & 0 & 0 & 1 & 0 \end{bmatrix}$$

$$-1 \cdot R_2 + R_1 \colon R_1 \begin{bmatrix} 0 & \frac{7}{4} & 1 & -\frac{1}{4} & 0 & \frac{5}{2} \\ 1 & \frac{1}{4} & 0 & \frac{1}{4} & 0 & \frac{1}{2} \\ -6 & -4 & 0 & 0 & 1 & 0 \end{bmatrix}$$

$$6 \cdot R_2 + R_3 \colon R_3 \begin{bmatrix} 0 & \frac{7}{4} & 1 & -\frac{1}{4} & 0 & \frac{5}{2} \\ 1 & \frac{1}{4} & 0 & \frac{1}{4} & 0 & \frac{1}{2} \\ 0 & -\frac{5}{2} & 0 & \frac{3}{2} & 1 & 3 \end{bmatrix}$$

5c. $(x_1,\ x_2,\ s_1,\ s_2) = \left(\frac{1}{2},\ 0,\ \frac{5}{2},\ 0\right),\ z = 3$

9. Let $x_1 =$ no. of loaves of bread; $x_2 =$ no. of cakes. Constraints:

$$\begin{aligned} 50x_1 + 30x_2 &\le 2400, \\ 0.90x_1 + 1.50x_2 &\le 190, \\ x_1,\ x_2 &\ge 0. \end{aligned}$$

Objective function is profit, to be maximized:

$$z = 1.20x_1 + 4.00x_2.$$

Convert constraints to equations and rearrange objective function:

$$\begin{aligned} 50x_1 + 30x_2 + s_1 + 0s_2 + 0z &= 2400, \\ 0.90x_1 + 1.50x_2 + 0s_1 + s_2 + 0z &= 190, \\ -1.20x_1 - 4.00x_2 + 0s_1 + 0s_2 + z &= 0, \end{aligned}$$

where s_1 is the slack variable for unused time and s_2 is the slack variable for unused money. The associated matrix is:

$$\begin{bmatrix} 50 & 30 & 1 & 0 & 0 & 2400 \\ 0.9 & 1.5 & 0 & 1 & 0 & 190 \\ -1.2 & -4 & 0 & 0 & 1 & 0 \end{bmatrix}.$$

Possible solution: $(x_1, x_2, s_1, s_2) = (0, 0, 2400, 190)$, $z = 0$. The pivot entry is in row 1, column 2. After pivoting, we obtain the matrix:

$$\begin{bmatrix} 1.667 & 1 & 0.033 & 0 & 0 & 80 \\ -1.6 & 0 & -0.05 & 1 & 0 & 70 \\ 5.467 & 0 & 0.133 & 0 & 1 & 320 \end{bmatrix}.$$

Since there are no negative entries in the objective row (row 3), this represents the maximal solution: $(x_1, x_2, s_1, s_2) = (0, 80, 0, 70)$, $z = 320$.

They should produce no bread and 80 cakes per day to realize the maximum profit of $320; this leaves no unused time and $70 of unused money.

13. Let x_1 = liters of House White; x_2 = liters of Premium White; x_3 = liters of Sauvignon Blanc. Constraints:

$$\begin{aligned} 0.75(2)x_1 + 0.25(2)x_2 &\leq 30,000, \\ 0.25(2)x_1 + 0.75(2)x_2 + 2x_3 &\leq 20,000, \\ x_1,\ x_2,\ x_3 &\geq 0. \end{aligned}$$

(Note the factors of 2; there are 2 lbs. of grapes used per liter.) Objective function is profit, to be maximized:

$$z = x_1 + 1.5x_2 + 2x_3.$$

Convert constraints to equations and rearrange objective function:

$$\begin{aligned} 1.5x_1 + 0.5x_2 + 0x_3 + s_1 + 0s_2 + 0z &= 30,000, \\ 0.5x_1 + 1.5x_2 + 2x_3 + 0s_1 + s_2 + 0z &= 20,000, \\ -x_1 - 1.5x_2 - 2x_3 + 0s_1 + 0s_2 + z &= 0, \end{aligned}$$

where s_1 is the slack variable for unused French colombard grapes and s_2 is the slack variable for unused sauvignon blanc grapes. The associated matrix is:

$$\begin{bmatrix} 1.5 & 0.5 & 0 & 1 & 0 & 0 & 30,000 \\ 0.5 & 1.5 & 2 & 0 & 1 & 0 & 20,000 \\ -1 & -1.5 & -2 & 0 & 0 & 1 & 0 \end{bmatrix}.$$

Possible solution: $(x_1, x_2, x_3, s_1, s_2) = (0, 0, 0, 30,000, 20,000)$, $z = 0$. The pivot entry is in row 2, column 3. After pivoting, we obtain the matrix:

$$\begin{bmatrix} 1.5 & 0.5 & 0 & 1 & 0 & 0 & 30,000 \\ 0.25 & 0.75 & 1 & 0 & 0.5 & 0 & 10,000 \\ -0.5 & 0 & 0 & 0 & 1 & 1 & 20,000 \end{bmatrix}.$$

Possible solution: $(x_1, x_2, x_3, s_1, s_2) = (0, 0, 10,000, 30,000, 0)$, $z = 20,000$. The pivot entry is in row 1, column 1. After pivoting, we obtain the matrix:

$$\begin{bmatrix} 1 & 0.333 & 0 & 0.667 & 0 & 0 & 20,000 \\ 0 & 0.667 & 1 & -0.167 & 0.5 & 0 & 5000 \\ 0 & 0.167 & 0 & 0.333 & 1 & 1 & 30,000 \end{bmatrix}.$$

Since there are no negative entries in the objective row (row 3), this represents the maximal solution: $(x_1, x_2, x_3, s_1, s_2) = (20,000,\ 0, 5000,\ 0,\ 0)$, $z = 30,000$. They should produce 20,000 liters of House White, no Premium White, and 5000 liters of Sauvignon Blanc this season to realize the maximum profit of $30,000; this uses up all of both grape varieties.

17. George Dantzig

21. This occurred when the objective function was rewritten so that all of the variables are on one side of the equation.

25. Associated matrix:

$$\begin{bmatrix} 72 & 46 & 73 & 26 & 54 & 1 & 0 & 0 & 185 \\ 37 & 84 & 45 & 83 & 85 & 0 & 1 & 0 & 237 \\ -17 & -26 & -85 & -63 & -43 & 0 & 0 & 1 & 0 \end{bmatrix}$$

Final simplex matrix:

$$\begin{bmatrix} 1.026 & 0.334 & 1 & 0 & 0.465 & 0.017 & -0.005 & 0 & 1.880 \\ -0.110 & 0.831 & 0 & 1 & 0.772 & -0.009 & 0.015 & 0 & 1.836 \\ 63.228 & 54.752 & 0 & 0 & 45.146 & 0.863 & 0.489 & 1 & 275.495 \end{bmatrix}$$

Solution: $(x_1, x_2, x_3, x_4, x_5, s_1, s_2) = (0,\ 0,\ 1.880,\ 1.836,\ 0,\ 0,\ 0)$, $z = 275.495$.

29. Let x_1 = no. of floor lamps, x_2 = no. of table lamps, and x_3 = no. of table lamps produced per week; s_1 = min. of time unused in wood shop, s_2 = min. of time unused in metal shop, s_3 = min. of time unused in electrical shop, and s_4 = min. of time unused in testing. Associated matrix:

$$\begin{bmatrix} 20 & 25 & 30 & 1 & 0 & 0 & 0 & 0 & 16,800 \\ 30 & 15 & 10 & 0 & 1 & 0 & 0 & 0 & 12,000 \\ 15 & 12 & 11 & 0 & 0 & 1 & 0 & 0 & 7200 \\ 5 & 5 & 4 & 0 & 0 & 0 & 1 & 0 & 2400 \\ -55 & -45 & -40 & 0 & 0 & 0 & 0 & 1 & 0 \end{bmatrix}$$

Final simplex matrix:

$$\begin{bmatrix} 0 & -10 & 0 & 1 & 1 & 0 & -10 & 0 & 4800 \\ 1 & 0.14 & 0 & 0 & 0.06 & 0 & -0.14 & 0 & 342.86 \\ 0 & -1.93 & 0 & 0 & -0.07 & 1 & -2.57 & 0 & 171.43 \\ 0 & 1.07 & 1 & 0 & -0.07 & 0 & 0.43 & 0 & 171.43 \\ 0 & 5.71 & 0 & 0 & 0.29 & 0 & 9.29 & 1 & 25,714.29 \end{bmatrix}$$

Solution: $(x_1, x_2, x_3, s_1, s_2, s_3, s_4) = (342.9,\ 0,\ 171.4,\ 4800,\ 0,\ 171.4,\ 0)$, $z = 25,714.29$.

33. Let

$x_1 =$ liters of House Red,
$x_2 =$ liters of Premium Red,
$x_3 =$ liters of Cabernet Sauvignon,
$x_4 =$ liters of Zinfandel,
$s_1 =$ pounds of unused pinot noir grapes,
$s_2 =$ pounds of unused zinfandel grapes,
$s_3 =$ pounds of unused gamay grapes,
$s_4 =$ pounds of unused cabernet sauvignon grapes.

We make a table of the data; the entries for grape quantities have been adjusted by multiplying each percentage by 2, since there are 2 lbs. of grapes needed per liter of wine:

	Pinot noir	Zinfandel	Gamay	Cabernet sauvignon	Profit
x_1	0.4	0.6	1	0	0.90
x_2	0.4	0	0.4	1.2	1.60
x_3	0	0	0	2	3.00
x_4	0	1.7	0.3	0	2.25

Associated matrix:

$$\begin{bmatrix} 0.4 & 0.4 & 0 & 0 & 1 & 0 & 0 & 0 & 0 & 30,000 \\ 0.6 & 0 & 0 & 1.7 & 0 & 1 & 0 & 0 & 0 & 30,000 \\ 1 & 0.4 & 0 & 0.3 & 0 & 0 & 1 & 0 & 0 & 30,000 \\ 0 & 1.2 & 2 & 0 & 0 & 0 & 0 & 1 & 0 & 22,000 \\ -0.9 & -1.6 & -3 & -2.25 & 0 & 0 & 0 & 0 & 1 & 0 \end{bmatrix}$$

Final simplex matrix:

$$\begin{bmatrix} 0 & 0.22 & 0 & 0 & 1 & 0.08 & -0.45 & 0 & 0 & 18,947.37 \\ 0 & -0.16 & 0 & 1 & 0 & 0.66 & -0.39 & 0 & 0 & 7894.74 \\ 1 & 0.45 & 0 & 0 & 0 & -0.20 & 1.12 & 0 & 0 & 27,631.58 \\ 0 & 0.60 & 1 & 0 & 0 & 0 & 0 & 0.50 & 0 & 11,000 \\ 0 & 0.25 & 0 & 0 & 0 & 1.30 & 0.12 & 1.50 & 1 & 75,631.58 \end{bmatrix}$$

Solutions: $(x_1, x_2, x_3, x_4, s_1, s_2, s_3, s_4) = (27,631.58, 0, 11,000, 7894.74, 18,947.37, 0, 0, 0)$, $z = 75,631.58$.

Mixed Constraints and Minimization

1. Minimizing w is equivalent to maximizing $-w = z$:

$$z = -2x_1 - 7x_2$$

Constraints:

$$\begin{aligned} x_1 + 4x_2 + s_1 + 0s_2 + 0s_3 + 0z &= 40, \\ 4x_1 + x_2 + 0s_1 + s_2 + 0s_3 + 0z &= 40, \\ x_1 + x_2 + 0s_1 + 0s_2 - s_3 + 0z &= 5, \end{aligned}$$

where s_1, s_2 are slack variables, and s_3 is a surplus variable; furthermore, s_1, s_2, $s_3 \geq 0$. The first simplex matrix:

$$\begin{bmatrix} 1 & 4 & 1 & 0 & 0 & 0 & 40 \\ 4 & 1 & 0 & 1 & 0 & 0 & 40 \\ 1 & 1 & 0 & 0 & -1 & 0 & 5 \\ 2 & 7 & 0 & 0 & 0 & 1 & 0 \end{bmatrix}$$

Since the corresponding solution has $s_3 = -5$ (column 5), we have a contradiction, hence this is not a possible solution. We pivot on row 3, column 2, to get column 2 into the form

$$\begin{bmatrix} 0 \\ 0 \\ 1 \\ 0 \end{bmatrix}.$$

$$\begin{bmatrix} -3 & 0 & 1 & 0 & 4 & 0 & 20 \\ 3 & 0 & 0 & 1 & 1 & 0 & 35 \\ 1 & 1 & 0 & 0 & -1 & 0 & 5 \\ -5 & 0 & 0 & 0 & 7 & 1 & -35 \end{bmatrix}$$

This matrix represents a possible solution, so we are now in phase II since the objective function row has negative values. Proceeding with the simplex method and pivoting as usual, we obtain the final simplex matrix:

$$\begin{bmatrix} 0 & 3 & 1 & 0 & 1 & 0 & 35 \\ 0 & -3 & 0 & 1 & 4 & 0 & 20 \\ 1 & 1 & 0 & 0 & -1 & 0 & 5 \\ 0 & 5 & 0 & 0 & 2 & 1 & -10 \end{bmatrix}$$

The optimal solution (maximal for z, minimal for w) is $(x_1, x_2, s_1, s_2, s_3) = (5, 0, 35, 20, 0)$, $z = -10$ or $-z = -(-w) = w = -(-10) = 10$.

5. Constraints:

$$\begin{aligned} 3x_1 - 2x_2 - s_1 + 0s_2 + 0s_3 + 0s_4 + 0z &= 25, \\ 3x_1 + x_2 + 0s_1 + s_2 + 0s_3 + 0s_4 + 0z &= 145, \\ x_1 + 0x_2 + 0s_1 + 0s_2 - s_3 + 0s_4 + 0z &= 20, \\ 0x_1 + x_2 + 0s_1 + 0s_2 + 0s_3 + s_4 + 0z &= 38, \end{aligned}$$

where s_2, $s_4 \geq 0$ are slack variables and s_1, $s_3 \geq 0$ are surplus variables. The first simplex matrix:

$$\begin{bmatrix} 3 & -2 & -1 & 0 & 0 & 0 & 0 & 25 \\ 3 & 1 & 0 & 1 & 0 & 0 & 0 & 145 \\ 1 & 0 & 0 & 0 & -1 & 0 & 0 & 20 \\ 0 & 1 & 0 & 0 & 0 & 1 & 0 & 38 \\ -5.2 & -1.3 & 0 & 0 & 0 & 0 & 1 & 0 \end{bmatrix}$$

Its corresponding solution is $(x_1, x_2, s_1, s_2, s_3, s_4) = (0, 0, -25, 145, -20, 38)$, $z = 0$. This is not a possible solution. We will eliminate the negative value of s_3 by changing the x_1 column into

$$\begin{bmatrix} 0 \\ 0 \\ 1 \\ 0 \\ 0 \end{bmatrix}.$$

After applying the operations $-3R_3 + R_1 : R_1$, $-3R_3 + R_2 : R_2$, $5.2R_3 + R_5 : R_5, -R_1 : R_1$, we obtain the matrix

$$\begin{bmatrix} 0 & 2 & 1 & 0 & -3 & 0 & 0 & 35 \\ 0 & 1 & 0 & 1 & 3 & 0 & 0 & 85 \\ 1 & 0 & 0 & 0 & -1 & 0 & 0 & 20 \\ 0 & 1 & 0 & 0 & 0 & 1 & 0 & 38 \\ 0 & -1.3 & 0 & 0 & -5.2 & 0 & 1 & 104 \end{bmatrix}$$

The corresponding solution is $(x_1, x_2, s_1, s_2, s_3, s_4) = (20, 0, 35, 85, 0, 38)$, $z = 104$. This matrix represents a possible solution, so we are now in phase II since the objective function row has negative values. Proceeding with the simplex method and pivoting as usual, we obtain the final simplex matrix:

$$\begin{bmatrix} 0 & 3 & 1 & 1 & 0 & 0 & 0 & 120 \\ 0 & 0.33 & 0 & 0.33 & 1 & 0 & 0 & 28.33 \\ 1 & 0.33 & 0 & 0.33 & 0 & 0 & 0 & 48.33 \\ 0 & 1 & 0 & 0 & 0 & 1 & 0 & 38 \\ 0 & 0.43 & 0 & 1.73 & 0 & 0 & 1 & 251.33 \end{bmatrix}$$

Solution: $(x_1, x_2, s_1, s_2, s_3, s_4) = (48.33, 0, 120, 0, 28.33, 38)$, $z = 251.33$. After doing all of this work, it helps to see what is going on if you now graph the constraints using graphing software, and then graph the objective function such that $z = 200$, $z = 300$, and $z = 251.33$.

9. Let $x_1 =$ lbs. whole sprouts, $x_2 =$ lbs. sprouts pieces, and $x_3 =$ lbs. sprouts salsa. Constraints:

$$\begin{aligned} -0.25x_1 + 0x_2 + x_3 - s_1 + 0s_2 + 0s_3 + 0s_4 + 0z &= 0, \\ x_1 + 0x_2 + 0x_3 + 0s_1 - s_2 + 0s_3 + 0s_4 + 0z &= 400,000, \\ 0x_1 + x_2 + 0x_3 + 0s_1 + 0s_2 - s_3 + 0s_4 + 0z &= 90,000, \\ x_1 + x_2 + x_3 + 0s_1 + 0s_2 + 0s_3 - s_4 + 0z &= 400,000 + 90,000 + 200,000 \\ &= 690,000, \end{aligned}$$

where s_1 is the number of pounds by which the sales of sprouts salsa exceeds a quarter of the sales of whole sprouts, s_2 is the number of pounds of whole sprouts sold over 400,000 lbs., s_3 is the number of pounds of sprouts pieces sold over 90,000 lbs., and s_4 is the number of pounds of sprouts products purchased over 690,000 lbs. We need to minimize the cost $w = 1.25x_1 + 1.75x_2 + 3x_3$, which is equivalent to maximizing $-w = z = -1.25x_1 - 1.75x_2 - 3x_3$. The first matrix is

$$\begin{bmatrix} -0.25 & 0 & 1 & -1 & 0 & 0 & 0 & 0 & 0 \\ 1 & 0 & 0 & 0 & -1 & 0 & 0 & 0 & 400,000 \\ 0 & 1 & 0 & 0 & 0 & -1 & 0 & 0 & 90,000 \\ 1 & 1 & 1 & 0 & 0 & 0 & -1 & 0 & 690,000 \\ 1.25 & 1.75 & 3 & 0 & 0 & 0 & 0 & 1 & 0 \end{bmatrix}.$$

Its corresponding solution is $(x_1, x_2, x_3, s_1, s_2, s_3, s_4) = (0, 0, 0, 0, -400,000, -90,000, -690,000)$, $z = 0$. This is not a possible solution. Suggested pivoting order: Phase I pivot on the element in (row, column) in this order: (4, 1), (3, 2), (2, 3), and then perform the operation $-R_1 : R_1$. We obtain the matrix

$$\begin{bmatrix} 0 & 0 & 0 & 1 & 1.25 & 1 & -1 & 0 & 100,000 \\ 0 & 0 & 1 & 0 & 1 & 1 & -1 & 0 & 200,000 \\ 0 & 1 & 0 & 0 & 0 & -1 & 0 & 0 & 90,000 \\ 1 & 0 & 0 & 0 & -1 & 0 & 0 & 0 & 400,000 \\ 0 & 0 & 0 & 0 & -1.75 & -1.25 & 3 & 1 & -1,257,500 \end{bmatrix}.$$

Its corresponding solution is $(x_1, x_2, x_3, s_1, s_2, s_3, s_4) = (400,000,\ 90,000,\ 200,000,\ 100,000,\ 0,\ 0,\ 0)$, $w = 1,257,500$. This is a possible solution, so we are now in phase II. The final simplex matrix is

$$\begin{bmatrix} 0 & 0 & 0 & 0.8 & 1 & 0.8 & -0.8 & 0 & 80,000 \\ 0 & 0 & 1 & -0.8 & 0 & 0.2 & -0.2 & 0 & 120,000 \\ 0 & 1 & 0 & 0 & 0 & -1 & 0 & 0 & 90,000 \\ 1 & 0 & 0 & 0.8 & 0 & 0.8 & -0.8 & 0 & 480,000 \\ 0 & 0 & 0 & 1.4 & 0 & 0.15 & 1.6 & 1 & -1,117,500 \end{bmatrix}.$$

Its corresponding solution is $(x_1, x_2, x_3, s_1, s_2, s_3, s_4) = (480,000,\ 90,000,\ 120,000,\ 0,\ 80,000,\ 0,\ 0)$, $w = 1,117,500$. In order to minimize production costs, the company should produce 480,000 lbs. of whole sprouts, 90,000 lbs. of sprouts pieces, and 120,000 lbs. of sprouts salsa.

13. *Hint:* In a minimization we use $z = -w$; in a maximization we use z. All variables are on one side of the equation.

17. Refer to the table. Let x_1 = millions of gallons of water supplied from the south reservoir to Crockett; x_2 = millions of gallons of water supplied from the south reservoir to Valona; x_3 = millions of gallons of water supplied from the north reservoir to Crockett; x_4 = millions of gallons of water supplied from the north reservoir to Valona. The south reservoir can supply up to 20 million gallons/day: $x_1 + x_2 \le 20 \Rightarrow x_1 + x_2 + s_1 = 20$. The north reservoir can supply up to 40 million gallons/day: $x_3 + x_4 \le 40 \Rightarrow x_3 + x_4 + s_2 = 40$. Crockett needs at least 30 million gallons/day: $x_1 + x_3 \ge 30 \Rightarrow x_1 + x_3 - s_3 = 30$. Valona needs at least 10 million gallons/day: $x_2 + x_4 \ge 10 \Rightarrow x_2 + x_4 - s_4 = 10$. Minimize $w = 7000x_1 + 9000x_2 + 11,000x_3 + 10,000x_4$; equivalently, maximize $-w = z = -7000x_1 - 9000x_2 - 11,000x_3 - 10,000x_4$. The first simplex matrix:

$$\begin{bmatrix} 1 & 1 & 0 & 0 & 1 & 0 & 0 & 0 & 0 & 20 \\ 0 & 0 & 1 & 1 & 0 & 1 & 0 & 0 & 0 & 40 \\ 1 & 0 & 1 & 0 & 0 & 0 & -1 & 0 & 0 & 30 \\ 0 & 1 & 0 & 1 & 0 & 0 & 0 & -1 & 0 & 10 \\ 7000 & 9000 & 11,000 & 10,000 & 0 & 0 & 0 & 0 & 1 & 0 \end{bmatrix}$$

Try this pivot order on elements in (row, column): Phase I pivots on (3, 3), (4,

2); Phase II pivot on (1, 1), (4, 4). You should get the final simplex matrix:

$$\begin{bmatrix} 1 & 1 & 0 & 0 & 1 & 0 & 0 & 0 & 0 & 20 \\ 0 & 0 & 0 & 0 & 1 & 1 & 1 & 1 & 0 & 20 \\ 0 & -1 & 1 & 0 & -1 & 0 & -1 & 0 & 0 & 10 \\ 0 & 1 & 0 & 1 & 0 & 0 & 0 & -1 & 0 & 10 \\ 0 & 3000 & 0 & 0 & 4000 & 0 & 11,000 & 10,000 & 1 & -350,000 \end{bmatrix}$$

1a. Let x_1 = no. of chairs produced per day and x_2 = no. of sofas produced per day. Constraints:

$$\begin{aligned} 6x_1 + 3x_2 + s_1 + 0s_2 + 0s_3 + 0z &= 96, \\ x_1 + x_2 + 0s_1 + s_2 + 0s_3 + 0z &= 18, \\ 2x_1 + 6x_2 + 0s_1 + 0s_2 + s_3 + 0z &= 72, \end{aligned}$$

where s_1 is a slack variable measuring hours of unused carpentry shop time, s_2 is a slack variable measuring hours of unused finishing shop time, and s_3 is a slack variable measuring hours of unused upholstery shop time; also $z = 160x_1 + 140x_2$ is the profit which must be maximized. The first simplex matrix:

$$\begin{bmatrix} 6 & 3 & 1 & 0 & 0 & 0 & 96 \\ 1 & 1 & 0 & 1 & 0 & 0 & 18 \\ 2 & 6 & 0 & 0 & 1 & 0 & 72 \\ -160 & -140 & 0 & 0 & 0 & 1 & 0 \end{bmatrix}$$

Pivot (phase I) on the elements in row 1, column 1, and then row 2, column 2. Final simplex matrix:

$$\begin{bmatrix} 1 & 0 & 0.3333 & -1 & 0 & 0 & 14 \\ 0 & 1 & -0.3333 & 2 & 0 & 0 & 4 \\ 0 & 0 & 1.3333 & -10 & 1 & 0 & 20 \\ 0 & 0 & 6.6667 & 120 & 0 & 1 & 2800 \end{bmatrix}$$

To maximize profit at \$2800.00, the factory should produce 14 chairs and 4 sofas per day.

1b. Using the final simplex matrix in 1a, note that the pertinent column for carpentry time (s_1) is column 3. If one extra hour of carpentry time is available, chair production will increase to $14 + 0.3333 = 14.3333$, sofa production will decrease to $4 - 0.3333 = 3.6667$, and profit will increase to $\$2800.00 + \$6.67 = \$2806.67$.

1c. Using the final simplex matrix in 1a, note that the pertinent column for finishing time (s_2) is column 4. If one extra hour of finishing time is available, chair production will decrease to $14 - 1 = 13$, sofa production will increase to $4 + 2 = 6$, and profit will increase to $\$2800.00 + \$120.00 = 2920.00$.

1d. Using the final simplex matrix in 1a, note that the pertinent column for upholstery time (s_3) is column 5. If one extra hour of upholstery time is available, daily production and profit will stay the same.

1e. From 1b-1d we observe that the most profit will be made by increasing the finishing shop time, thus if the factory were to hire another worker, it should be a finishing worker. The factory will still profit if it pays this worker less than \$120.00 per hour.

1f. It should consider retraining one of the upholstery shop workers, since the shop has the least effect on productivity and profits and the most surplus time.

5a. Coffee tables, as seen in the s_1-column in the final simplex matrix of the original craftsman problem from sec. 3.2.

5b. 0.25 tables per day (row 2, column 3).

5c. $\frac{2.5}{0.25} = 10$.

5d. Since, for every hour increase in time spent there is a decrease in coffee table production, this type of profit increase can be sustained until coffee table production drops to zero. That occurs at 10 such hour increases (from 5c), or an increase of 10 hours.

9a. Recall from Sec. 3.2, Exercise 9, that x_1 = no. of loaves of bread; x_2 = no. of cakes, and the constraints were

$$\begin{aligned} 50x_1 + 30x_2 + s_1 + 0s_2 + 0z &= 2400, \\ 0.90x_1 + 1.50x_2 + 0s_1 + s_2 + 0z &= 190, \\ -1.20x_1 - 4.00x_2 + 0s_1 + 0s_2 + z &= 0, \end{aligned}$$

where s_1 is the slack variable for unused time and s_2 is the slack variable for unused money. The final simplex matrix was $\begin{bmatrix} 1.667 & 1 & 0.033 & 0 & 0 & 80 \\ -1.6 & 0 & -0.05 & 1 & 0 & 70 \\ 5.467 & 0 & 0.133 & 0 & 1 & 320 \end{bmatrix}$. The pertinent column for labor is column 3. If 1 extra hour of labor becomes available, the number of loaves of bread produced would decrease to $0 - 0.05(60) = -3$, an impossibility. (Note: the factor of 60 is needed to convert -0.05, which is a decrease per *minute*, to a decrease per hour. Remember that the unit of time we used in this solution was in minutes.) Thus we recompute with the entry 2460 (minutes, or 41 hours) in the first simplex matrix in row 1, column 6:

$$\begin{bmatrix} 50 & 30 & 1 & 0 & 0 & 2460 \\ 0.9 & 1.5 & 0 & 1 & 0 & 190 \\ -1.2 & -4 & 0 & 0 & 1 & 0 \end{bmatrix}$$

Final simplex matrix:

$$\begin{bmatrix} 1.667 & 1 & 0.033 & 0 & 0 & 82 \\ -1.6 & 0 & -0.05 & 1 & 0 & 67 \\ 5.467 & 0 & 0.133 & 0 & 1 & 328 \end{bmatrix}$$

We observe that production of cakes increases to 82, production of loaves of bread remains at zero, and profit increases to \$328.00. Thus in cases where we appear to get negative values when production of an item was originally zero, we can ignore that and use the shortcuts discussed in the text, viz., an extra $0.033\ldots$ loaves per minute (or $60 \cdot 0.033\ldots = 2$ loaves per hour), and an extra $60 \cdot 0.133\ldots = \$8.00$.

9b. $\frac{70}{0.05} = 1400$ min. $= 23.33$ hrs. more OK.

9c. First simplex matrix:

$$\begin{bmatrix} 50 & 30 & 1 & 0 & 0 & 2400 \\ 0.9 & 1.5 & 0 & 1 & 0 & 191 \\ -1.2 & -4 & 0 & 0 & 1 & 0 \end{bmatrix}$$

Final simplex matrix:

$$\begin{bmatrix} 1.667 & 1 & 0.033 & 0 & 0 & 80 \\ -1.6 & 0 & -0.05 & 1 & 0 & 71 \\ 5.467 & 0 & 0.133 & 0 & 1 & 320 \end{bmatrix}$$

There is no change in productivity, hence no change in profit. Or, using the shortcut from the text, x_2 changes from 80 to $80 + 0 = 80$ and z from 320 to $320 + 0 = 320$. In other words, neither variable changes.

9d. Indeterminate.

Duality

1. We first make a matrix of the primal model:

$$\begin{bmatrix} 8 & 4 & 9 & 12 \\ 7 & 11 & 913 & 48 \\ 5 & 3 & 4 & 0 \end{bmatrix}$$

Then we alter the matrix by changing its rows into columns, i.e., row 1 becomes column 1, row 2 becomes column 2, and row 3 becomes column 3. This new matrix is called the *transpose* of the original matrix, and is usually denoted with a superscript "T". Thus if the first matrix is labelled A, the transpose would be labelled A^T.

$$\begin{bmatrix} 8 & 7 & 5 \\ 4 & 11 & 3 \\ 9 & 913 & 4 \\ 12 & 48 & 0 \end{bmatrix}$$

Now we convert the transpose into the dual model. Constraints:

$$8y_1 + 7y_2 \geq 5,$$
$$4y_1 + 11y_2 \geq 3,$$
$$9y_1 + 913y_2 \geq 4.$$

Minimize $w = 12y_1 + 48y_2$.

5a. Minimum is $z = 1120$; $y_1 = 0$; $y_2 = 9.7$.

5b. Maximum is $z = 1120$; $x_1 = 0$ (shadow value of r_1); $x_2 = 0.33$ (shadow value of r_2).

9. Matrix of the primal model:

$$\begin{bmatrix} 15 & 1 & 25 \\ 1 & 6 & 20 \\ 8 & 7 & 78 \\ 2 & 3 & 0 \end{bmatrix}$$

Matrix of the dual model:

$$\begin{bmatrix} 15 & 1 & 8 & 2 \\ 1 & 6 & 7 & 3 \\ 25 & 20 & 78 & 0 \end{bmatrix}$$

Maximize $z = 25x_1 + 20x_2 + 78x_3$ subject to the constraints

$$15x_1 + x_2 + 8x_3 \leq 2,$$
$$x_1 + 6x_2 + 7x_3 \leq 3.$$

First simplex matrix:

$$\begin{bmatrix} 15 & 1 & 8 & 1 & 0 & 0 & 2 \\ 1 & 6 & 7 & 0 & 1 & 0 & 3 \\ -25 & -20 & -78 & 0 & 0 & 1 & 0 \end{bmatrix}$$

Final simplex matrix:

$$\begin{bmatrix} 2.171 & 0 & 1 & 0.146 & -0.024 & 0 & 0.220 \\ -2.366 & 1 & 0 & -0.171 & 0.195 & 0 & 0.244 \\ 97 & 0 & 0 & 8 & 2 & 1 & 22 \end{bmatrix}$$

$y_1 = 8$ (shadow value of s_1, column 4), $y_2 = 2$ (shadow value of s_2, column 5), $z = 22$.

13. Let $y_1 =$ oz. of chili macs, $y_2 =$ oz. of whipped potatoes. Minimize the objective cost function $w = 15y_1 + 2y_2$ subject to the constraints

$$92y_1 + 13y_2 \geq 3500,$$
$$5.8y_1 + 0.4y_2 \geq 120,$$
$$221y_1 + 72y_2 \geq 750.$$

Matrix of the primal model:

$$\begin{bmatrix} 92 & 13 & 3500 \\ 5.8 & 0.4 & 120 \\ 221 & 72 & 750 \\ 15 & 2 & 0 \end{bmatrix}$$

Matrix of the dual model:

$$\begin{bmatrix} 92 & 5.8 & 221 & 15 \\ 13 & 0.4 & 72 & 2 \\ 3500 & 120 & 750 & 0 \end{bmatrix}$$

Dual model: Maximize $z = 3500x_1 + 120x_2 + 750x_3$ subject to the constraints

$$92x_1 + 5.8x_2 + 221x_3 \leq 15,$$
$$13x_1 + 0.4x_2 + 72x_3 \leq 2.$$

First simplex matrix:

$$\begin{bmatrix} 92 & 5.8 & 221 & 1 & 0 & 0 & 15 \\ 13 & 0.4 & 72 & 0 & 1 & 0 & 2 \\ -3500 & -120 & -750 & 0 & 0 & 1 & 0 \end{bmatrix}$$

Final simplex matrix (all entries rounded to 3 decimal places):

$$\begin{bmatrix} 0 & 1 & -97.176 & 0.337 & -2.383 & 0 & 0.285 \\ 1 & 0 & 8.528 & -0.010 & 0.150 & 0 & 0.145 \\ 0 & 0 & 17,438.601 & 4.145 & 239.896 & 1 & 541.969 \end{bmatrix}$$

Thus $y_1 = 4.145$, $y_2 = 239.896$, and the minimum cost is 542¢ = \$5.42 per meal.

1a. Let x_1 = no. of model 110 speaker assemblies and x_2 = no. of model 330 speaker assemblies. Maximize $z = 200x_1 + 350x_2$ subject to the constraints

$$\begin{aligned} x_1 + 2x_2 &\le 90, \\ x_1 + x_2 &\le 60, \\ 0x_1 + x_2 &\le 44. \end{aligned}$$

First simplex matrix:

$$\begin{bmatrix} 1 & 2 & 1 & 0 & 0 & 0 & 90 \\ 1 & 1 & 0 & 1 & 0 & 0 & 60 \\ 0 & 1 & 0 & 0 & 1 & 0 & 44 \\ -200 & -350 & 0 & 0 & 0 & 1 & 0 \end{bmatrix}$$

Final simplex matrix:

$$\begin{bmatrix} 1 & 0 & -1 & 2 & 0 & 0 & 30 \\ 0 & 0 & -1 & 1 & 1 & 0 & 14 \\ 0 & 1 & 1 & -1 & 0 & 0 & 30 \\ 0 & 0 & 150 & 50 & 0 & 1 & 16,500 \end{bmatrix}$$

They should make 30 model 110 assemblies and 30 model 330 assemblies.

1b. Maximum income is $16,500.00.

1c. If they have one more tweeter in stock, their income will increase by $150.00 (row 4, column 3, final simplex matrix); also, production of model 110 assemblies will decrease by 1 (row 1, column 3) and production of model 330 assemblies will increase by 1 (row 3, column 3).

1d. If they have one more midrange speaker in stock, their income will increase by $50.00 (row 4, column 4, final simplex matrix); also, production of model 110 assemblies will increase by 2 (row 1, column 4) and production of model 330 assemblies will decrease by 1 (row 3, column 4).

1e. If they have one more woofer in stock, their income will increase by $0.00 (row 4, column 5, final simplex matrix); also, production of model 110 assemblies will remain the same (row 1, column 5) and production of model 330 assemblies will remain the same (row 3, column 5).

2. Let x_1 = no. of model 110 speaker assemblies, x_2 = no. of model 220 speaker assemblies, and x_3 = no. of model 330 speaker assemblies. Maximize $z = 200x_1 + + 280x_2 + 350x_3$ subject to the constraints

$$\begin{aligned} x_1 + x_2 + 2x_3 &\le 140, \\ x_1 + x_2 + x_3 &\le 90, \\ 0x_1 + x_2 + x_3 &\le 66. \end{aligned}$$

First simplex matrix:

$$\begin{bmatrix} 1 & 1 & 2 & 1 & 0 & 0 & 0 & 140 \\ 1 & 1 & 1 & 0 & 1 & 0 & 0 & 90 \\ 0 & 1 & 1 & 0 & 0 & 1 & 0 & 66 \\ -200 & -280 & -350 & 0 & 0 & 0 & 1 & 0 \end{bmatrix}$$

Final simplex matrix:

$$\begin{bmatrix} 1 & 0 & 0 & 0 & 1 & -1 & 0 & 24 \\ 0 & 1 & 0 & -1 & 1 & 1 & 0 & 16 \\ 0 & 0 & 1 & 1 & -1 & 0 & 0 & 50 \\ 0 & 0 & 0 & 70 & 130 & 80 & 1 & 26,780 \end{bmatrix}$$

They should make 24 model 110 assemblies, 16 model 220 assemblies, and 50 model 330 assemblies. Each slack variable (columns 4 to 6) is zero, which means that all of the speakers in stock will be used.

3a. Let $x_1 =$ no. of Sunnys stocked and $x_2 =$ no. of Iwas stocked. We want to maximize the profit objective function $z = 220x_1 + 200x_2$ subject to the constraints

$$\begin{aligned} 12x_1 + 8x_2 &\leq 12,000, \\ x_1 + x_2 &\geq 600, \\ x_1 &\geq 2x_2. \end{aligned}$$

We insert the slack variable s_1 for empty storage space and the surplus variables s_2 for the number of personal stereos sold over 600 and s_3 for the number of Sunnys sold over twice the number of Iwas sold:

$$\begin{aligned} 12x_1 + 8x_2 + s_1 + 0s_2 + 0s_3 + 0z &= 12,000, \\ x_1 + x_2 + 0s_1 - s_2 + 0s_3 + 0z &= 600, \\ x_1 - 2x_2 + 0s_1 + 0s_2 - s_3 + 0z &= 0. \end{aligned}$$

First simplex matrix:

$$\begin{bmatrix} 12 & 8 & 1 & 0 & 0 & 0 & 12,000 \\ 1 & 1 & 0 & -1 & 0 & 0 & 600 \\ 1 & -2 & 0 & 0 & -1 & 0 & 0 \\ -220 & -200 & 0 & 0 & 0 & 1 & 0 \end{bmatrix}$$

After Phase I pivots on the elements in row 2, column 2, and then row 3, column 1, you should get the matrix

$$\begin{bmatrix} 0 & 0 & 1 & \frac{32}{3} & \frac{4}{3} & 0 & 5600 \\ 0 & 1 & 0 & -\frac{1}{3} & \frac{1}{3} & 0 & 200 \\ 1 & 0 & 0 & -\frac{2}{3} & -\frac{1}{3} & 0 & 400 \\ 0 & 0 & 0 & -\frac{640}{3} & -\frac{20}{3} & 1 & 128,000 \end{bmatrix}.$$

Now proceed with Phase II pivoting and obtain the final simplex matrix

$$\begin{bmatrix} 0 & 0 & 0.09375 & 1 & 0.12500 & 0 & 525 \\ 0 & 1 & 0.03125 & 0 & 0.37500 & 0 & 375 \\ 1 & 0 & 0.06250 & 0 & -0.25000 & 0 & 750 \\ 0 & 0 & 20 & 0 & 20 & 1 & 240,000 \end{bmatrix}.$$

They should stock 750 Sunnys and 375 Iwas.

3b. Maximum profit is $240,000.00.

3c. Column 3 (s_1) has to do with storage space. Note from the final simplex matrix that 1 extra cubic foot of storage space will yield a \$20 profit increase. However, the stocks of Sunnys and Iwas would increase by fractional amounts, which is not possible. In order to make this realistic, a space increase of $1/0.0625 = 16$ ft^3 will make room for 1 more Sunny and $16(0.03125) = \frac{1}{2}$ Iwa. Thus an increase of 32 ft^3 will make room for 2 Sunnys and an Iwa, and the profit would increase to $\$240{,}000 + 32(\$20) = \$240{,}640.00$.

4a. Let y_1 = ounces of Chinchilla Vanilla and y_2 = ounces of Science Chinchilla. Minimize $w = 32y_1 + 40y_2$; equivalently, maximize $-w = z = -32y_1 - 40y_2$ subject to the constraints

$$\begin{aligned} 2y_1 + 5y_2 &\geq 8, \\ 2y_1 + 3y_2 &\geq 5, \\ y_1 + y_2 &\geq 2. \end{aligned}$$

First simplex matrix:

$$\begin{bmatrix} 2 & 5 & -1 & 0 & 0 & 0 & 8 \\ 2 & 3 & 0 & -1 & 0 & 0 & 5 \\ 1 & 1 & 0 & 0 & -1 & 0 & 2 \\ 32 & 40 & 0 & 0 & 0 & 1 & 0 \end{bmatrix}$$

After Phase I pivots on the entries in row 3, column 1, then row 2, column 2, and finally row 1, column 4, you should get the final simplex matrix

$$\begin{bmatrix} 0 & 0 & -\frac{1}{3} & 1 & -\frac{4}{3} & 0 & \frac{1}{3} \\ 0 & 1 & -\frac{1}{3} & 0 & \frac{2}{3} & 0 & \frac{4}{3} \\ 1 & 0 & \frac{1}{3} & 0 & -\frac{5}{3} & 0 & \frac{2}{3} \\ 0 & 0 & \frac{8}{3} & 0 & \frac{80}{3} & 1 & -\frac{224}{3} \end{bmatrix}.$$

Thus, $(y_1, y_2, r_1, r_2, r_3) = \left(\frac{2}{3}, \frac{4}{3}, 0, \frac{1}{3}, 0\right)$, and $w = \frac{224}{3} \approx 75$. To achieve the minimum cost of \$0.75 per day, feed $\frac{2}{3}$ oz. of Chinchilla Vanilla and $\frac{4}{3}$ oz. of Science Chinchilla. This allows no excess protein nor vitamins and allows an extra $\frac{1}{3}$ unit of carbohydrates.

4b. Primal matrix:

$$\begin{bmatrix} 2 & 5 & 8 \\ 2 & 3 & 5 \\ 1 & 1 & 2 \\ 32 & 40 & 0 \end{bmatrix}$$

Dual matrix:

$$\begin{bmatrix} 2 & 2 & 1 & 32 \\ 5 & 3 & 1 & 40 \\ 8 & 5 & 2 & 0 \end{bmatrix}$$

First simplex matrix:

$$\begin{bmatrix} 2 & 2 & 1 & 1 & 0 & 0 & 32 \\ 5 & 3 & 1 & 0 & 1 & 0 & 40 \\ -8 & -5 & -2 & 0 & 0 & 1 & 0 \end{bmatrix}$$

Final simplex matrix:

$$\begin{bmatrix} 0 & \frac{4}{3} & 1 & \frac{5}{3} & -\frac{2}{3} & 0 & \frac{80}{3} \\ 1 & \frac{1}{3} & 0 & -\frac{1}{3} & \frac{1}{3} & 0 & \frac{8}{3} \\ 0 & \frac{1}{3} & 0 & \frac{2}{3} & \frac{4}{3} & 1 & \frac{224}{3} \end{bmatrix}$$

5. George Dantzig

Chapter 4 Matrix Equations

Section 4.0 Matrix Arithmetic

1a. $M+N=\begin{bmatrix}3&7\\2&1\end{bmatrix}+\begin{bmatrix}5&-2\\0&3\end{bmatrix}=\begin{bmatrix}3+5&7+(-2)\\2+0&1+3\end{bmatrix}=\begin{bmatrix}8&5\\2&4\end{bmatrix}$

1b. $N+M=\begin{bmatrix}5&-2\\0&3\end{bmatrix}+\begin{bmatrix}3&7\\2&1\end{bmatrix}=\begin{bmatrix}5+3&-2+7\\0+2&3+1\end{bmatrix}=\begin{bmatrix}8&5\\2&4\end{bmatrix}$

1c. Matrix addition is commutative because addition of real numbers is commutative.

5a. $B+G=\begin{bmatrix}8&0\\-3&10\end{bmatrix}$

5b.

$$\begin{aligned}2G+3B&=2\begin{bmatrix}1&0\\-2&8\end{bmatrix}+3\begin{bmatrix}7&0\\-1&2\end{bmatrix}\\&=\begin{bmatrix}2\cdot1&2\cdot0\\2(-2)&2\cdot8\end{bmatrix}+\begin{bmatrix}3\cdot7&3\cdot0\\3(-1)&3\cdot2\end{bmatrix}\\&=\begin{bmatrix}2&0\\-4&16\end{bmatrix}+\begin{bmatrix}21&0\\-3&6\end{bmatrix}\\&=\begin{bmatrix}23&0\\-7&22\end{bmatrix}\end{aligned}$$

5c.

$$\begin{aligned}BG&=\begin{bmatrix}7&0\\-1&2\end{bmatrix}\begin{bmatrix}1&0\\-2&8\end{bmatrix}\\&=\begin{bmatrix}7\cdot1+0(-2)&7\cdot0+0\cdot8\\-1(1)+2(-2)&-1\cdot0+2\cdot8\end{bmatrix}\\&=\begin{bmatrix}7&0\\-5&16\end{bmatrix}\end{aligned}$$

5d.

$$\begin{aligned} GB &= \begin{bmatrix} 1 & 0 \\ -2 & 8 \end{bmatrix} \begin{bmatrix} 7 & 0 \\ -1 & 2 \end{bmatrix} \\ &= \begin{bmatrix} 1 \cdot 7 + 0(-1) & 1 \cdot 0 + 0 \cdot 2 \\ -2 \cdot 7 + 8(-1) & -2 \cdot 0 + 8 \cdot 2 \end{bmatrix} \\ &= \begin{bmatrix} 7 & 0 \\ -22 & 16 \end{bmatrix} \end{aligned}$$

9a. Does not exist (different dimensions).

9b.

$$\begin{aligned} AB &= \begin{bmatrix} 3 & 1 \\ 4 & 6 \\ 8 & 2 \end{bmatrix} \begin{bmatrix} 7 & 0 \\ -1 & 2 \end{bmatrix} \\ &= \begin{bmatrix} 3 \cdot 7 + 1(-1) & 3 \cdot 0 + 1 \cdot 2 \\ 4 \cdot 7 + 6(-1) & 4 \cdot 0 + 6 \cdot 2 \\ 8 \cdot 7 + 2(-1) & 8 \cdot 0 + 2 \cdot 2 \end{bmatrix} \\ &= \begin{bmatrix} 20 & 2 \\ 22 & 12 \\ 54 & 4 \end{bmatrix} \end{aligned}$$

9c. Does not exist.

13a.

$$\begin{aligned} F(B+G) &= [5 \quad 2] \left(\begin{bmatrix} 7 & 0 \\ -1 & 2 \end{bmatrix} + \begin{bmatrix} 1 & 0 \\ -2 & 8 \end{bmatrix} \right) \\ &= [5 \quad 2] \begin{bmatrix} 8 & 0 \\ -3 & 10 \end{bmatrix} \\ &= [5 \cdot 8 + 2(-3) \quad 5 \cdot 0 + 2 \cdot 10] \\ &= [34 \quad 20] \end{aligned}$$

13b.

$$\begin{aligned} FB + FG &= [5 \quad 2] \begin{bmatrix} 7 & 0 \\ -1 & 2 \end{bmatrix} + [5 \quad 2] \begin{bmatrix} 1 & 0 \\ -2 & 8 \end{bmatrix} \\ &= [5 \cdot 7 + 2(-1) \quad 5 \cdot 0 + 2 \cdot 2] + [5 \cdot 1 + 2(-2) \quad 5 \cdot 0 + 2 \cdot 8] \\ &= [33 \quad 4] + [1 \quad 16] \\ &= [34 \quad 20] \end{aligned}$$

13c. Yes, since $F(B+G) = FB + FG$.

17.

$$\begin{aligned} & [2 \quad 1] \begin{bmatrix} 1.25 & 1.30 \\ 0.95 & 1.10 \end{bmatrix} \\ &= [2 \cdot 1.25 + 1 \cdot 0.95 \quad 2 \cdot 1.30 + 1 \cdot 1.10] \\ &= [3.45 \quad 3.70] \end{aligned}$$

21. First we convert the grade matrix into a grade point matrix:

$$\begin{bmatrix} 4 & 3 & 3 & 2 \\ 2 & 4 & 4 & 2 \\ 3 & 4 & 3 & 4 \end{bmatrix}$$

Row 1 represents Jose's grade points, row 2 represents Eloise's grade points, and row 3 represents Sylvie's grade points. Highest possible units $\times$ grade points $= 4 \cdot 4 + 4 \cdot 4 + 3 \cdot 4 + 3 \cdot 4 = 56$. For example, Jose's units $\times$ grade points $= 43$, and $43/56 \approx 0.768$. The highest possible units $\times$ grade points/ $56 = 56/56 = 1$. The GPA is found by multiplying (units $\times$ grade points / 56) by 4. In other words, the scalar is $4/56 = 1/14$. To get GPA's, do the following:

$$\frac{1}{14}\begin{bmatrix} 4 & 3 & 3 & 2 \\ 2 & 4 & 4 & 2 \\ 3 & 4 & 3 & 4 \end{bmatrix}\begin{bmatrix} 4 \\ 4 \\ 3 \\ 3 \end{bmatrix} = \frac{1}{14}\begin{bmatrix} 43 \\ 42 \\ 49 \end{bmatrix}$$
$$\approx \begin{bmatrix} 3.071 \\ 3.000 \\ 3.500 \end{bmatrix}$$

Thus Jose and Eloise have a B average and Sylvie has a B+ average.

25. $\begin{bmatrix} 1 & 0 \\ 0 & 1 \end{bmatrix} \cdot \begin{bmatrix} 3 & -2 \\ 4 & 0 \end{bmatrix} = \begin{bmatrix} 1 \cdot 3 + 0 \cdot 4 & 1(-2) + 0 \cdot 0 \\ 0 \cdot 3 + 1 \cdot 4 & 0(-2) + 1 \cdot 0 \end{bmatrix} = \begin{bmatrix} 3 & -2 \\ 4 & 0 \end{bmatrix}$

29.
$$\begin{bmatrix} 19 & 7 & 34 \\ 74 & 0 & -11 \\ 13 & -2 & 44 \end{bmatrix} \cdot \begin{bmatrix} 1 & 0 & 0 \\ 0 & 1 & 0 \\ 0 & 0 & 1 \end{bmatrix} = \begin{bmatrix} 19 \cdot 1 + 7 \cdot 0 + 34 \cdot 0 & 19 \cdot 0 + 7 \cdot 1 + 34 \cdot 0 & 19 \cdot 0 + 7 \cdot 0 + 34 \cdot 1 \\ 74 \cdot 1 + 0 \cdot 0 - 11 \cdot 0 & 74 \cdot 0 + 0 \cdot 1 - 11 \cdot 0 & 74 \cdot 0 + 0 \cdot 0 - 11 \cdot 1 \\ 13 \cdot 1 - 2 \cdot 0 + 44 \cdot 0 & 13 \cdot 0 - 2 \cdot 1 + 44 \cdot 0 & 13 \cdot 0 - 2 \cdot 0 + 44 \cdot 1 \end{bmatrix} =$$
$$\begin{bmatrix} 19 & 7 & 34 \\ 74 & 0 & -11 \\ 13 & -2 & 44 \end{bmatrix}$$

33. *Hint:* $X = \frac{1}{7}\left(\begin{bmatrix} 5 & 7 \\ 2 & 14 \end{bmatrix} + \begin{bmatrix} 6 & 3 \\ 2 & -4 \end{bmatrix}\right)$

49a, b.

$$N^2 = NN$$
$$= \begin{bmatrix} 0 & 1 & 0 & 0 & 0 \\ 0 & 0 & 1 & 0 & 0 \\ 1 & 0 & 0 & 1 & 0 \\ 0 & 0 & 0 & 0 & 1 \\ 0 & 0 & 1 & 0 & 0 \end{bmatrix}\begin{bmatrix} 0 & 1 & 0 & 0 & 0 \\ 0 & 0 & 1 & 0 & 0 \\ 1 & 0 & 0 & 1 & 0 \\ 0 & 0 & 0 & 0 & 1 \\ 0 & 0 & 1 & 0 & 0 \end{bmatrix}$$
$$= \begin{bmatrix} 0 & 0 & 1 & 0 & 0 \\ 1 & 0 & 0 & 1 & 0 \\ 0 & 1 & 0 & 0 & 1 \\ 0 & 0 & 1 & 0 & 0 \\ 1 & 0 & 0 & 1 & 0 \end{bmatrix}$$

This matrix shows routing which is possible through an intermediary city. For example, there is an indirect flight from San Francisco to Los Angeles (row 5, column 4) which is routed through the intermediary city of Denver--from San Francisco to Denver (row 5, column 3 of N) and then from Denver to Los Angeles (row 3, column 4 of N). The matrix N^3 would show routing which is possible through 2 intermediary cities, and in general, N^k (k a positive integer) shows routing which is possible through $k - 1$ intermediary cities.

49c.

$$N + N^2 = \begin{bmatrix} 0 & 1 & 1 & 0 & 0 \\ 1 & 0 & 1 & 1 & 0 \\ 1 & 1 & 0 & 1 & 1 \\ 0 & 0 & 1 & 0 & 1 \\ 1 & 0 & 1 & 1 & 0 \end{bmatrix}$$

49d. 1's represent a route that's either direct or with 1 stopover.

49e. $N + N^2 + N^3 + N^4 = \begin{bmatrix} 1 & 2 & 1 & 1 & 1 \\ 1 & 1 & 3 & 1 & 1 \\ 3 & 1 & 2 & 3 & 1 \\ 1 & 1 & 1 & 1 & 2 \\ 1 & 1 & 3 & 1 & 1 \end{bmatrix}$

This shows flights, either direct or indirect, between all of the cities given. Entries greater than 1 indicate more than one routing alternative for flights.

49f. Los Angeles to Chicago and San Francisco to Atlanta.

Section 4.1
Inverse Matrices

1.

$$\begin{aligned}
AB &= \begin{bmatrix} 7 & 1 \\ -3 & 2 \end{bmatrix} \begin{bmatrix} \frac{2}{17} & -\frac{1}{17} \\ \frac{3}{17} & \frac{7}{17} \end{bmatrix} \\
&= \begin{bmatrix} 7 & 1 \\ -3 & 2 \end{bmatrix} \left(\frac{1}{17} \begin{bmatrix} 2 & -1 \\ 3 & 7 \end{bmatrix} \right) \\
&= \left(\frac{1}{17} \begin{bmatrix} 7 & 1 \\ -3 & 2 \end{bmatrix} \right) \begin{bmatrix} 2 & -1 \\ 3 & 7 \end{bmatrix} \\
&= \frac{1}{17} \left(\begin{bmatrix} 7 & 1 \\ -3 & 2 \end{bmatrix} \begin{bmatrix} 2 & -1 \\ 3 & 7 \end{bmatrix} \right) \\
&= \frac{1}{17} \begin{bmatrix} 17 & 0 \\ 0 & 17 \end{bmatrix} \\
&= \begin{bmatrix} 1 & 0 \\ 0 & 1 \end{bmatrix}
\end{aligned}$$

They are inverses of each other.

5. *Pre*multiply both sides of the equation by $\begin{bmatrix} \frac{2}{17} & -\frac{1}{17} \\ \frac{3}{17} & \frac{7}{17} \end{bmatrix}$.

9. *Pre*multiply both sides of the equation by $\begin{bmatrix} -\frac{1}{27} & \frac{6}{27} \\ \frac{5}{27} & -\frac{3}{27} \end{bmatrix}$.

13a.

$$\begin{aligned}
\begin{bmatrix} 4 & -1 \\ 2 & -2 \end{bmatrix} \begin{bmatrix} x \\ y \end{bmatrix} &= \begin{bmatrix} 4x - y \\ 2x - 2y \end{bmatrix} \\
&= \begin{bmatrix} 3 \\ -1 \end{bmatrix},
\end{aligned}$$

whence we obtain the system

$$\begin{aligned}
4x - y &= 3, \\
2x - 2y &= -1.
\end{aligned}$$

17.

$$\begin{aligned}
\begin{bmatrix} 5 & -7 \\ 3 & 2 \end{bmatrix} \begin{bmatrix} x \\ y \end{bmatrix} &= \begin{bmatrix} 1 \\ -18 \end{bmatrix} \\
\begin{bmatrix} \frac{2}{31} & \frac{7}{31} \\ -\frac{3}{31} & \frac{5}{31} \end{bmatrix} \begin{bmatrix} 5 & -7 \\ 3 & 2 \end{bmatrix} \begin{bmatrix} x \\ y \end{bmatrix} &= \begin{bmatrix} \frac{2}{31} & \frac{7}{31} \\ -\frac{3}{31} & \frac{5}{31} \end{bmatrix} \begin{bmatrix} 1 \\ -18 \end{bmatrix} \\
\begin{bmatrix} 1 & 0 \\ 0 & 1 \end{bmatrix} \begin{bmatrix} x \\ y \end{bmatrix} &= \begin{bmatrix} -4 \\ -3 \end{bmatrix} \\
\begin{bmatrix} x \\ y \end{bmatrix} &= \begin{bmatrix} -4 \\ -3 \end{bmatrix}
\end{aligned}$$

21.

$$\begin{bmatrix} 9 & 2 & -1 \\ 3 & -2 & 4 \\ 1 & 0 & -1 \end{bmatrix} \begin{bmatrix} x \\ y \\ z \end{bmatrix} = \begin{bmatrix} 20 \\ -23 \\ 6 \end{bmatrix}$$

$$\begin{bmatrix} \frac{1}{15} & \frac{1}{15} & \frac{1}{5} \\ \frac{7}{30} & -\frac{4}{15} & -\frac{13}{10} \\ \frac{1}{15} & \frac{1}{15} & -\frac{4}{5} \end{bmatrix} \begin{bmatrix} 9 & 2 & -1 \\ 3 & -2 & 4 \\ 1 & 0 & -1 \end{bmatrix} \begin{bmatrix} x \\ y \\ z \end{bmatrix} = \begin{bmatrix} \frac{1}{15} & \frac{1}{15} & \frac{1}{5} \\ \frac{7}{30} & -\frac{4}{15} & -\frac{13}{10} \\ \frac{1}{15} & \frac{1}{15} & -\frac{4}{5} \end{bmatrix} \begin{bmatrix} 20 \\ -23 \\ 6 \end{bmatrix}$$

$$\begin{bmatrix} \mathbf{1} & 0 & 0 \\ 0 & \mathbf{1} & 0 \\ 0 & 0 & \mathbf{1} \end{bmatrix} \begin{bmatrix} x \\ y \\ z \end{bmatrix} = \begin{bmatrix} 1 \\ 3 \\ -5 \end{bmatrix}$$

$$\begin{bmatrix} x \\ y \\ z \end{bmatrix} = \begin{bmatrix} 1 \\ 3 \\ -5 \end{bmatrix}$$

25. Let $x =$ oz. of beef liver and $y =$ cups of Velveeta Surprise. We obtain the system

$$40x + 340y = 740,$$
$$1.4x + 14y = 28.$$

The equivalent matrix equation is

$$\begin{bmatrix} 40 & 340 \\ 1.4 & 14 \end{bmatrix} \begin{bmatrix} x \\ y \end{bmatrix} = \begin{bmatrix} 740 \\ 28 \end{bmatrix}.$$

Since $\begin{bmatrix} 40 & 340 \\ 1.4 & 14 \end{bmatrix}^{-1} = \begin{bmatrix} \frac{1}{6} & -\frac{85}{21} \\ -\frac{1}{60} & \frac{10}{21} \end{bmatrix}$ (the -1 superscript indicates the inverse), we use it to premultiply the matrix equation on both sides to solve the equation:

$$\begin{bmatrix} \frac{1}{6} & -\frac{85}{21} \\ -\frac{1}{60} & \frac{10}{21} \end{bmatrix} \begin{bmatrix} 40 & 340 \\ 1.4 & 14 \end{bmatrix} \begin{bmatrix} x \\ y \end{bmatrix} = \begin{bmatrix} \frac{1}{6} & -\frac{85}{21} \\ -\frac{1}{60} & \frac{10}{21} \end{bmatrix} \begin{bmatrix} 740 \\ 28 \end{bmatrix}$$

$$\begin{bmatrix} \mathbf{1} & 0 \\ 0 & \mathbf{1} \end{bmatrix} \begin{bmatrix} x \\ y \end{bmatrix} = \begin{bmatrix} 10 \\ 1 \end{bmatrix}$$

$$\begin{bmatrix} x \\ y \end{bmatrix} = \begin{bmatrix} 10 \\ 1 \end{bmatrix}$$

The patient will receive 10 ounces of beef liver and 1 cup of Velveeta Surprise each day. The patient envies the dogs in Exercise 26.

29a. (See the text for details on inverting this matrix.)

29b.

$$
\begin{aligned}
AA^{-1} &= \begin{bmatrix} a & b \\ c & d \end{bmatrix}\left(\frac{1}{da-bc}\begin{bmatrix} d & -b \\ -c & a \end{bmatrix}\right) \\
&= \begin{bmatrix} a & b \\ c & d \end{bmatrix}\begin{bmatrix} \frac{d}{da-bc} & \frac{-b}{da-bc} \\ \frac{-c}{da-bc} & \frac{a}{da-bc} \end{bmatrix} \\
&= \begin{bmatrix} \frac{ad}{da-bc}+\frac{-bc}{da-bc} & \frac{-ab}{da-bc}+\frac{ba}{da-bc} \\ \frac{cd}{da-bc}+\frac{-dc}{da-bc} & \frac{-cb}{da-bc}+\frac{da}{da-bc} \end{bmatrix} \\
&= \begin{bmatrix} \frac{da-bc}{da-bc} & \frac{-ab+ab}{da-bc} \\ \frac{cd-cd}{da-bc} & \frac{da-bc}{da-bc} \end{bmatrix} \\
&= \begin{bmatrix} 1 & 0 \\ 0 & 1 \end{bmatrix}
\end{aligned}
$$

29c.

$$
\begin{aligned}
A^{-1}A &= \frac{1}{da-bc}\begin{bmatrix} d & -b \\ -c & a \end{bmatrix}\begin{bmatrix} a & b \\ c & d \end{bmatrix} \\
&= \frac{1}{da-bc}\begin{bmatrix} da-bc & db-db \\ -ac+ac & -bc+da \end{bmatrix} \\
&= \begin{bmatrix} 1 & 0 \\ 0 & 1 \end{bmatrix}
\end{aligned}
$$

33. The number zero does not have a multiplicative inverse. For if it does, call it 0^{-1}. Then $0 \cdot 0^{-1} = 1$, a contradiction because zero times any number is zero. This is equivalent to saying that you can't divide by zero, because dividing by zero is the same as multiplying by 0^{-1}, an impossibility because 0^{-1} does not exist. (E.g., dividing by 5 is the same as multiplying by $5^{-1} = \frac{1}{5}$.) If a matrix has no inverse, one would get undefined entries when attempting to compute an inverse matrix.

37. $(x, y, z) \approx (-2.0059, 7.0799, 4.7626)$

39d. If the determinant of a matrix is zero, then the inverse of the matrix does not exist, and conversely.

Section 4.2 The Gauss-Jordan Method and Inverses

1.

$$\begin{bmatrix} 5 & 7 & -2 & 1 & 0 & 0 \\ 3 & -1 & 1 & 0 & 1 & 0 \\ 8 & 6 & 9 & 0 & 0 & 1 \end{bmatrix}$$

$$\frac{1}{5}R1 : R1 \begin{bmatrix} 1 & \frac{7}{5} & -\frac{2}{5} & \frac{1}{5} & 0 & 0 \\ 3 & -1 & 1 & 0 & 1 & 0 \\ 8 & 6 & 9 & 0 & 0 & 1 \end{bmatrix}$$

$$-3R1 + R2 : R2 \begin{bmatrix} 1 & \frac{7}{5} & -\frac{2}{5} & \frac{1}{5} & 0 & 0 \\ 0 & -\frac{26}{5} & \frac{11}{5} & -\frac{3}{5} & 1 & 0 \\ 8 & 6 & 9 & 0 & 0 & 1 \end{bmatrix}$$

$$-8R1 + R3 : R3 \begin{bmatrix} 1 & \frac{7}{5} & -\frac{2}{5} & \frac{1}{5} & 0 & 0 \\ 0 & -\frac{26}{5} & \frac{11}{5} & -\frac{3}{5} & 1 & 0 \\ 0 & -\frac{26}{5} & \frac{61}{5} & -\frac{8}{5} & 0 & 1 \end{bmatrix}$$

$$-\frac{5}{26}R2 : R2 \begin{bmatrix} 1 & \frac{7}{5} & -\frac{2}{5} & \frac{1}{5} & 0 & 0 \\ 0 & 1 & -\frac{11}{26} & \frac{3}{26} & -\frac{5}{26} & 0 \\ 0 & -\frac{26}{5} & \frac{61}{5} & -\frac{8}{5} & 0 & 1 \end{bmatrix}$$

$$-\frac{7}{5}R2 + R1 : R1 \begin{bmatrix} 1 & 0 & \frac{5}{26} & \frac{1}{26} & \frac{7}{26} & 0 \\ 0 & 1 & -\frac{11}{26} & \frac{3}{26} & -\frac{5}{26} & 0 \\ 0 & -\frac{26}{5} & \frac{61}{5} & -\frac{8}{5} & 0 & 1 \end{bmatrix}$$

$$\frac{26}{5}R2 + R3 : R3 \begin{bmatrix} 1 & 0 & \frac{5}{26} & \frac{1}{26} & \frac{7}{26} & 0 \\ 0 & 1 & -\frac{11}{26} & \frac{3}{26} & -\frac{5}{26} & 0 \\ 0 & 0 & 10 & -1 & -1 & 1 \end{bmatrix}$$

$$\frac{1}{10}R3 : R3 \begin{bmatrix} 1 & 0 & \frac{5}{26} & \frac{1}{26} & \frac{7}{26} & 0 \\ 0 & 1 & -\frac{11}{26} & \frac{3}{26} & -\frac{5}{26} & 0 \\ 0 & 0 & 1 & -\frac{1}{10} & -\frac{1}{10} & \frac{1}{10} \end{bmatrix}$$

$$\frac{11}{26}R3 + R2 : R2 \begin{bmatrix} 1 & 0 & \frac{5}{26} & \frac{1}{26} & \frac{7}{26} & 0 \\ 0 & 1 & 0 & \frac{19}{260} & -\frac{61}{260} & \frac{11}{260} \\ 0 & 0 & 1 & -\frac{1}{10} & -\frac{1}{10} & \frac{1}{10} \end{bmatrix}$$

$$-\frac{5}{26}R3 + R1 : R1 \begin{bmatrix} 1 & 0 & 0 & \frac{3}{52} & \frac{15}{52} & -\frac{1}{52} \\ 0 & 1 & 0 & \frac{19}{260} & -\frac{61}{260} & \frac{11}{260} \\ 0 & 0 & 1 & -\frac{1}{10} & -\frac{1}{10} & \frac{1}{10} \end{bmatrix}$$

Thus the inverse is $\begin{bmatrix} \frac{3}{52} & \frac{15}{52} & -\frac{1}{52} \\ \frac{19}{260} & -\frac{61}{260} & \frac{11}{260} \\ -\frac{1}{10} & -\frac{1}{10} & \frac{1}{10} \end{bmatrix}$.

5a. Premultiply both sides of the matrix equation

$$\begin{bmatrix} 5 & 7 & -2 \\ 3 & -1 & 1 \\ 8 & 6 & 9 \end{bmatrix} \begin{bmatrix} x \\ y \\ z \end{bmatrix} = \begin{bmatrix} 3 \\ 2 \\ 14 \end{bmatrix}$$

by the inverse matrix $\begin{bmatrix} \frac{3}{52} & \frac{15}{52} & -\frac{1}{52} \\ \frac{19}{260} & -\frac{61}{260} & \frac{11}{260} \\ -\frac{1}{10} & -\frac{1}{10} & \frac{1}{10} \end{bmatrix}$ (from Exercise 1) to obtain the solution $\begin{bmatrix} x \\ y \\ z \end{bmatrix} = \begin{bmatrix} \frac{25}{52} \\ \frac{89}{260} \\ \frac{9}{10} \end{bmatrix}$.

5b. Premultiply both sides of the matrix equation

$$\begin{bmatrix} 5 & 7 & -2 \\ 3 & -1 & 1 \\ 8 & 6 & 9 \end{bmatrix} \begin{bmatrix} x \\ y \\ z \end{bmatrix} = \begin{bmatrix} 12 \\ 11 \\ 19 \end{bmatrix}$$

by the inverse matrix $\begin{bmatrix} \frac{3}{52} & \frac{15}{52} & -\frac{1}{52} \\ \frac{19}{260} & -\frac{61}{260} & \frac{11}{260} \\ -\frac{1}{10} & -\frac{1}{10} & \frac{1}{10} \end{bmatrix}$ (from Exercise 1) to obtain the solution $\begin{bmatrix} x \\ y \\ z \end{bmatrix} = \begin{bmatrix} \frac{7}{2} \\ -\frac{9}{10} \\ -\frac{2}{5} \end{bmatrix}$.

9. $\begin{bmatrix} 3 & -1 & -1 \\ 2 & 7 & -8 \\ 5 & 1 & -3 \end{bmatrix}^{-1} = \frac{1}{28} \begin{bmatrix} -13 & -4 & 15 \\ -34 & -4 & 22 \\ -33 & -8 & 23 \end{bmatrix}$; $\begin{bmatrix} x \\ y \\ z \end{bmatrix} = \begin{bmatrix} \frac{83}{2} \\ 55 \\ \frac{115}{2} \end{bmatrix}$

Section 4.3
Leontiff Input-Output Models

1. We add to Robinson's and Friday's demand matrix $\begin{bmatrix} 1460 \\ 1460 \end{bmatrix}$ Robinson's wife's demand matrix $\begin{bmatrix} 1095 \\ 0 \end{bmatrix}$ to obtain the new demand matrix:

$$\begin{bmatrix} 1460 \\ 1460 \end{bmatrix} + \begin{bmatrix} 1095 \\ 0 \end{bmatrix} = \begin{bmatrix} 2555 \\ 1460 \end{bmatrix}.$$

The necessary production matrix is found by:

$$\begin{aligned} P &= (I-T)^{-1}D \\ &= \frac{1}{0.49}\begin{bmatrix} 1 & 5 \\ 0.1 & 0.99 \end{bmatrix}\begin{bmatrix} 2555 \\ 1460 \end{bmatrix} \\ &= \frac{1}{0.49}\begin{bmatrix} 9855 \\ 1700.9 \end{bmatrix} \\ &\approx \begin{bmatrix} 20,112 \\ 3471 \end{bmatrix} \end{aligned}$$

5a.

$$\begin{aligned} P &= (I-T)^{-1}D \\ &= \left(\begin{bmatrix} 1 & 0 \\ 0 & 1 \end{bmatrix} - \begin{bmatrix} 0.1 & 0.2 \\ 0.1 & 0.3 \end{bmatrix}\right)^{-1}\begin{bmatrix} 2000 \\ 3000 \end{bmatrix} \\ &= \begin{bmatrix} 0.9 & -0.2 \\ -0.1 & 0.7 \end{bmatrix}^{-1}\begin{bmatrix} 2000 \\ 3000 \end{bmatrix} \\ &= \frac{1}{0.61}\begin{bmatrix} 0.7 & 0.2 \\ 0.1 & 0.9 \end{bmatrix}\begin{bmatrix} 2000 \\ 3000 \end{bmatrix} \\ &= \frac{1}{0.61}\begin{bmatrix} 2000 \\ 2900 \end{bmatrix} \\ &\approx \begin{bmatrix} 3279 \\ 4754 \end{bmatrix} \end{aligned}$$

5b.

T :

Production of 1 unit of coal requires:	Production of 1 unit of steel requires:
0.1 units of coal	0.2 units of coal
0.1 units of steel	0.3 units of steel

The entries in the production matrix are the respective units of coal and steel needed to produce the respective units of coal and steel in the demand matrix.

9. Here are the relevant matrices: $T = \begin{bmatrix} 0.10 & 0.35 \\ 0.25 & 0.10 \end{bmatrix}$, $D = \begin{bmatrix} 25,000,000,000 \\ 25,000,000 \end{bmatrix}$.

Review Exercises

1. $AB = \begin{bmatrix} 71 & -11 \\ 10 & -32 \end{bmatrix}$

2. $BA = \begin{bmatrix} -15 & 26 \\ 52 & 54 \end{bmatrix}$

3. $CD = \begin{bmatrix} 20 & 5 \\ 16 & 4 \end{bmatrix}$

4. $DC = [24]$

5. $AC = \begin{bmatrix} 53 \\ 26 \end{bmatrix}$

6. Does not exist.

7. Does not exist.

8.

$$\begin{aligned} 5A - B &= 5\begin{bmatrix} 5 & 7 \\ 6 & -1 \end{bmatrix} - \begin{bmatrix} 3 & -5 \\ 8 & 2 \end{bmatrix} \\ &= \begin{bmatrix} 25 & 35 \\ 30 & -5 \end{bmatrix} - \begin{bmatrix} 3 & -5 \\ 8 & 2 \end{bmatrix} \\ &= \begin{bmatrix} 22 & 40 \\ 22 & -7 \end{bmatrix} \end{aligned}$$

9. $|A| = 5(-1) - 6 \cdot 7 = -47$

10.

$$\begin{aligned} A^{-1}B &= \begin{bmatrix} 5 & 7 \\ 6 & -1 \end{bmatrix}^{-1} \begin{bmatrix} 3 & -5 \\ 8 & 2 \end{bmatrix} \\ &= \frac{1}{47}\begin{bmatrix} 1 & 7 \\ 6 & -5 \end{bmatrix}\begin{bmatrix} 3 & -5 \\ 8 & 2 \end{bmatrix} \\ &\approx \begin{bmatrix} 1.2553 & 0.1915 \\ -0.4681 & -0.8511 \end{bmatrix} \end{aligned}$$

11. Yes, $A + (B + C) = (A + B) + C$ where addition is defined.

12a. No, $A - B \neq B - A$. (For example, use matrices A and B given for Problems 1-10.)

12b. No, $A - (B - C) \neq (A - B) - C$. (For example, use matrices A and B (twice!) given for Problems 1-10.)

13a. No, $AB \neq BA$. (For example, use matrices A and B given for Problems 1-10.)

13b. Yes, $(AB)C = A(BC)$ where multiplication is defined.

13c. Yes, $A(B+C) = AB + AC$ where addition and multiplication are defined.

14. $\frac{1}{7}\begin{bmatrix} 4 & 3 \\ 2 & 4 \end{bmatrix}\begin{bmatrix} 4 \\ 3 \end{bmatrix} \approx \begin{bmatrix} 3.571 \\ 2.857 \end{bmatrix}$. Thus Al has a B+ average and Tipper has a C+ average.

17a.

N :

	H	**L**	**P**	**S**	**T**
H	0	0	0	0	1
L	0	0	0	1	0
P	0	0	0	1	0
S	0	1	1	0	1
T	1	0	0	1	0

17b.

$$N^2 = \begin{bmatrix} 1 & 0 & 0 & 1 & 0 \\ 0 & 1 & 1 & 0 & 1 \\ 0 & 1 & 1 & 0 & 1 \\ 1 & 0 & 0 & 3 & 0 \\ 0 & 1 & 1 & 0 & 2 \end{bmatrix}$$

This represents indirect connections between cities with one intervening stop. If an entry is $n > 1$, that means that there are n different ways to accomplish the indirect flight between two given cities.

17c.

$$N + N^2 = \begin{bmatrix} 1 & 0 & 0 & 1 & 1 \\ 0 & 1 & 1 & 1 & 1 \\ 0 & 1 & 1 & 1 & 1 \\ 1 & 1 & 1 & 3 & 1 \\ 1 & 1 & 1 & 1 & 2 \end{bmatrix}$$

This represents direct connections or indirect connections with one intervening stop between cities.

17d.

$$N + N^2 + N^3 = \begin{bmatrix} 1 & 1 & 1 & 1 & 3 \\ 1 & 1 & 1 & 4 & 1 \\ 1 & 1 & 1 & 4 & 1 \\ 1 & 4 & 4 & 3 & 5 \\ 3 & 1 & 1 & 5 & 2 \end{bmatrix}$$

This represents connections between every given city; one can go from one city to any other city with a maximum of two stops. In some cases, there is more than one way to accomplish this.

17e. Hong Kong to Los Angeles and Hong Kong to Portland.

18.

$$A^{-1} = \begin{bmatrix} \frac{2}{5} & -\frac{67}{15} & -\frac{14}{15} \\ -\frac{1}{5} & \frac{17}{5} & \frac{4}{5} \\ \frac{1}{5} & -\frac{26}{15} & -\frac{7}{15} \end{bmatrix}$$

21. First we must convert the given encrypted sequence into a matrix. What is its size? To decrypt the matrix, the encrypted matrix must be premultiplied by the matrix E^{-1}, where $E = \begin{bmatrix} 4 & 3 & 1 \\ 2 & -1 & 5 \\ 8 & 0 & -2 \end{bmatrix}$, which is a 3×3 matrix. Thus in order for the multiplication to be possible, the encrypted matrix must have the size $3 \times n$. In other words, it must have 3 rows and n columns. Since there are 21 numbers in the encrypted sequence, we will create a 3×7 encrypted matrix:

$$\begin{bmatrix} 88 & 119 & 144 & 162 & 141 & 147 & 70 \\ 128 & 101 & 72 & 126 & 99 & 169 & 124 \\ -22 & 160 & 66 & 62 & 84 & 168 & -46 \end{bmatrix}$$

Now, $E^{-1} = \begin{bmatrix} \frac{1}{74} & \frac{3}{74} & \frac{4}{37} \\ \frac{11}{37} & -\frac{4}{37} & -\frac{9}{74} \\ \frac{2}{37} & \frac{6}{37} & -\frac{5}{74} \end{bmatrix}$, and

$$\begin{bmatrix} \frac{1}{74} & \frac{3}{74} & \frac{4}{37} \\ \frac{11}{37} & -\frac{4}{37} & -\frac{9}{74} \\ \frac{2}{37} & \frac{6}{37} & -\frac{5}{74} \end{bmatrix} \begin{bmatrix} 88 & 119 & 144 & 162 & 141 & 147 & 70 \\ 128 & 101 & 72 & 126 & 99 & 169 & 124 \\ -22 & 160 & 66 & 62 & 84 & 168 & -46 \end{bmatrix}$$
$$= \begin{bmatrix} 4 & 23 & 12 & 14 & 15 & 27 & 1 \\ 15 & 5 & 27 & 27 & 21 & 5 & 13 \\ 27 & 12 & 15 & 25 & 18 & 24 & 27 \end{bmatrix}$$

Now we decrypt:

4	15	27	23	5	12	12
D	O		W	E	L	L
27	15	14	27	25	15	21
	O	N		Y	O	U
18	27	5	24	1	13	27
R		E	X	A	M	

"DO WELL ON YOUR EXAM"

22. The amount consumed in the production process, TP, is subtracted from the original amount produced, P, to yield the demand D which shows the part of the production which can be used to satisfy the consumers' demand.

23.

$$\begin{aligned} P - TP &= D \\ IP - TP &= D \\ (I - T)P &= D \\ (I - T)^{-1}(I - T)P &= (I - T)^{-1}D \\ IP &= (I - T)^{-1}D \\ P &= (I - T)^{-1}D \end{aligned}$$

This formula shows the quantities which must be produced (i.e., the production matrix P) given the technology matrix T and the demand matrix D. This is a common problem in input-output modelling.

25. Cayley and Sylvester

26. Not all matrices have inverses. The ones which do not are precisely those with zero determinant.

27. Wassily Leontieff

Chapter 5
Sets

Section 5.1
Sets and Set Operations

1a. This is well defined, provided we know what is meant by an automobile! For example, are we including trucks as well as cars? Is an SUV an automobile?

1b. This is not well defined because "inexpensive" is a matter of opinion. (A recent development called *fuzzy set theory* is intended to be able to handle sets like these.)

1c. This is well defined because a prime number is defined according to precise mathematical rules.

1d. This is not well defined because a "large" number is a matter of opinion.

5. S, {yes, no}, {yes, undecided}, {no, undecided}, {yes}, {no}, {undecided}, $\emptyset$. All of the above are proper subsets except for the improper subset S.

9. $B' =$ {Monday, Tuesday, Wednesday, Thursday}

13. $A \cap B$

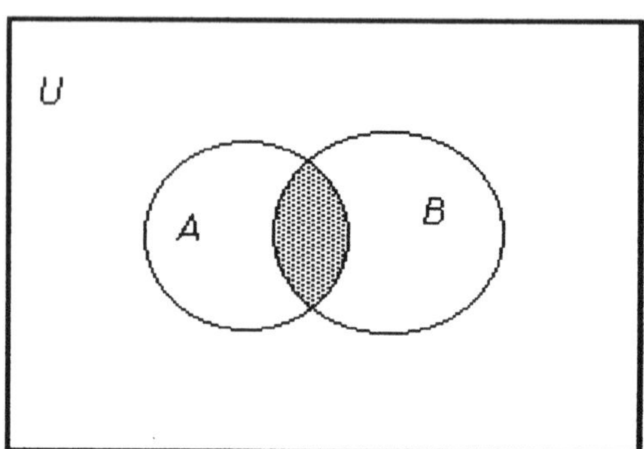

17. $A \cup B'$

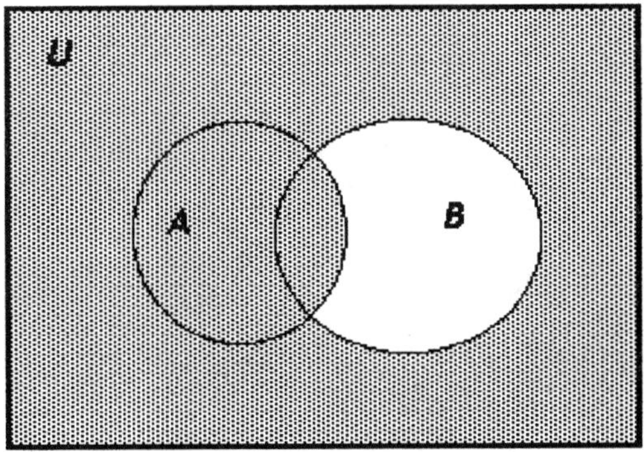

21. $A' \cup B'$

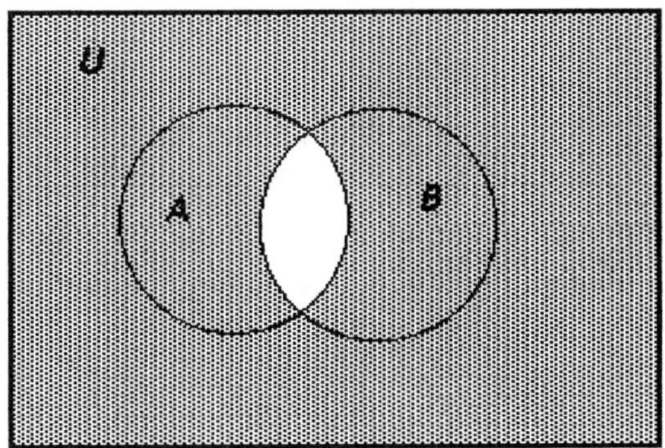

23a.

$$
\begin{aligned}
n(A \cup B) - n(A \cup B) &= n(A) + n(B) - n(A \cap B) - n(A \cup B) \\
0 + n(A \cap B) &= n(A) + n(B) - n(A \cap B) - n(A \cup B) + n(A \cap B) \\
n(A \cap B) &= n(A) + n(B) - n(A \cup B) \\
&= 37 + 84 - 100 \\
&= 21
\end{aligned}
$$

This means that set A has 37 elements, set B has 84 elements, and they have 21 elements in common. Thus set A has $37 - 21 = 16$ elements not in B and set B has $84 - 21 = 63$ elements not in A. Since U has 150 elements and $A \cup B$ has

100 elements, there are 50 elements in U which are not in $A \cup B$.

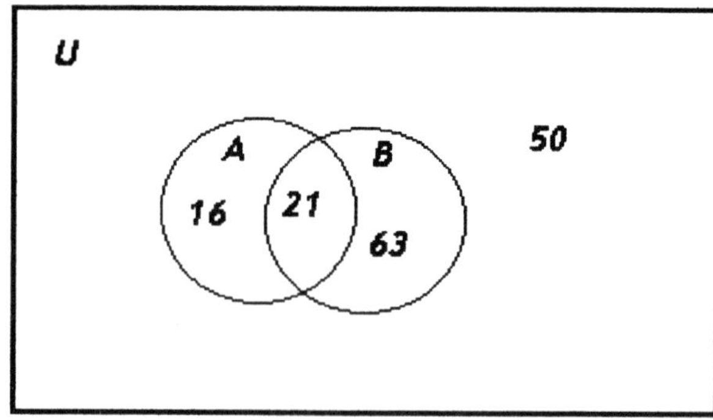

23b.

$$\begin{aligned} n(A \cap B) &= n(A) + n(B) - n(A \cup B) \\ &= 37 + 84 - 121 \\ &= 0 \end{aligned}$$

The intersection of A and B is empty, which means that the sets are *disjoint*.

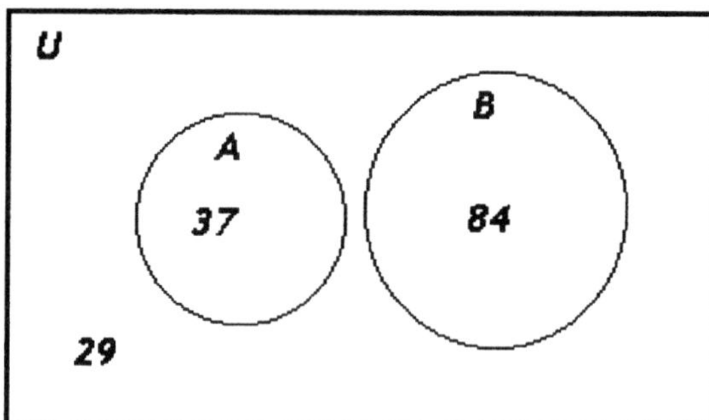

25a. Let A equal the set of students who checked the automobile box and B equal the set of students who checked the motorcycle box. There may be some rich students who own both, which means that it is possible that $A \cap B \neq \emptyset$. Indeed, we are given that 29 students checked both boxes. Now,

$$\begin{aligned} n(A \cup B) &= n(A) + n(B) - n(A \cap B) \\ &= 91 + 123 - 29 \\ &= 185 \end{aligned}$$

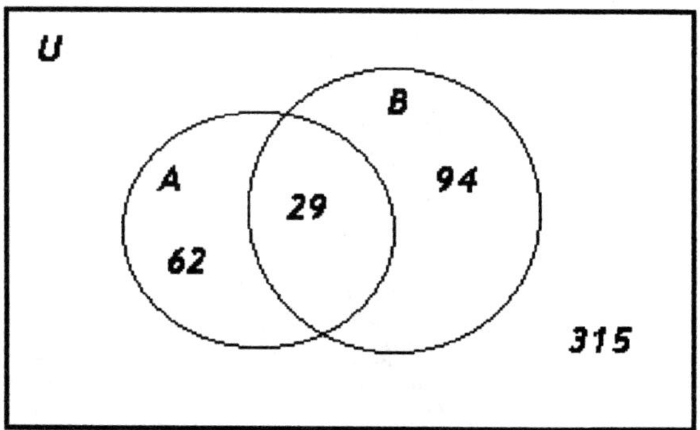

25b. The percent of students who own an automobile or a motorcycle (inclusive) is $\frac{n(A \cup B)}{n(U)} \cdot 100 = \frac{185}{500} \cdot 100 = 0.37 \cdot 100 = 37.$

Note: Use the following table for Exercises 29-31:

U.S. STATES
Alabama
Alaska
Arizona
Arkansas
California
Colorado
Connecticut
Delaware
Florida
Georgia
Hawaii
Idaho
Illinois
Indiana
Iowa
Kansas
Kentucky
Louisiana
Maine
Maryland
Massachusetts
Michigan
Minnesota
Mississippi
Missouri
Montana
Nebraska
Nevada
New Hampshire
New Jersey
New Mexico
New York
North Carolina
North Dakota
Ohio
Oklahoma
Oregon
Pennsylvania
Rhode Island
South Carolina
South Dakota
Tennessee
Texas
Utah
Vermont
Virginia
Washington
West Virginia
Wisconsin
Wyoming

29. $n(M') = n(U) - n(M) = 50 - 8 = 42$

Note: For Exercises 33-35 use the following table:

MONTHS
April
August
December
February
January
July
June
March
May
November
October
September

33. $n(R') = n(U) - n(R) = 12 - 4 = 8$

Note: For Exercises 37-45 use the following table of a deck of 52 standard playing cards; there are four *suits* (spades, clubs, hearts, and diamonds) with 13 cards each:

BLACK		RED	
♠ SPADES	♣ CLUBS	♥ HEARTS	♦ DIAMONDS
Ace	Ace	Ace	Ace
King	King	King	King
Queen	Queen	Queen	Queen
Jack	Jack	Jack	Jack
10	10	10	10
9	9	9	9
8	8	8	8
7	7	7	7
6	6	6	6
5	5	5	5
4	4	4	4
3	3	3	3
2	2	2	2

Note also that when interpreting "word" problems such as the following (in the context of set theory), the word "or" generally means *union* and the word "and" generally means *intersection*.

37. There are 13 spades and 4 Aces, and one Ace of spades; thus the number of spades or Aces is $13+4-1=16$.

39. *Face* cards are the ones with "faces," namely, the King, Queen, and Jack. There are 12 face cards and 26 black cards, and 6 black face cards. Thus the number of face cards or black cards is $12+26-6=32$.

41. There are 6 cards which are both face cards and black cards.

45. No card is both an Ace and an 8.

49a. $\{a\}=A$, $\emptyset$. Two subsets

49b. $\{a,b\}=A$, $\{a\}$, $\{b\}$, $\emptyset$. Four subsets

49c. $\{a,b,c\}=A$, $\{a,b\}$, $\{a,c\}$, $\{b,c\}$, $\{a\}$, $\{b\}$, $\{c\}$, $\emptyset$. Eight subsets

49d. $\{a,b,c,d\}=A$, $\{a,b,c\}$, $\{a,b,d\}$, $\{a,c,d\}$, $\{b,c,d\}$, $\{a,b\}$, $\{a,c\}$, $\{a,d\}$, $\{b,c\}$, $\{b,d\}$, $\{c,d\}$, $\{a\}$, $\{b\}$, $\{c\}$, $\{d\}$, $\emptyset$. Sixteen subsets

49e. Yes; if the cardinal number of set A is n, then A has 2^n subsets.

49f. Since $n(A)=6$, A has $2^6=64$ subsets.

53. The set $\{0\}$ has one element, viz. 0, whereas the set $\emptyset=\{\ \}$ has no elements.

57. It is advantageous to use the roster method when the number of its elements is finite; otherwise set builder notation can be used. Also, set builder notation is useful when the elements can be specified with a function.

61. If N is in John's group, then P is in Juneko's group. But J and P must not be in the same group, so J (selection a) is not in Juneko's group.

Section 5.2
Applications of Venn Diagrams

1a. Let $V = \{\text{people}|\text{the person owned a VCR}\}$, $M = \{\text{people}|\text{the person owned a microwave oven}\}$. We have that $n(V) = 94$, $n(M) = 127$, and $n(V \cap M) = 78$. The number of people who owned a VCR or a microwave oven is

$$\begin{aligned} n(V \cup M) &= n(V) + n(M) - n(V \cap M) \\ &= 94 + 127 - 78 \\ &= 143 \end{aligned}$$

1b. Since 94 people owned a VCR and 78 people owned a VCR and a microwave oven, it follows that $94 - 78 = 16$ people owned a VCR but not a microwave oven.

1c. Since 127 people owned a microwave oven and 78 people owned a a microwave oven and a VCR, it follows that $127 - 78 = 49$ people owned a microwave oven but not a VCR.

1d. Since 200 people were surveyed and $n(V \cup M) = 143$, it follows that the number of people who owned neither a VCR nor a microwave oven is $n((V \cup M)') = n(U) - n(V \cup M) = 200 - 143 = 57$.

5a. Since 289 adults watched the New Movie, and of those, 183 didn't watch the Big Game, then $289 - 183 = 106$ watched both.

5b. Since 674 adults were surveyed, and of those, 226 watched neither program, then $674 - 226 = 448$ watched at least one program.

5c. Recall that 106 people watched both programs, 289 watched the New Movie, and 448 watched at least one program. The number of adults who watched *only* the Big Game was $448 - 289 = 159$, whence $159 + 106 = 265$ watched the Big Game.

5d. 159 (see Exercise 5c)

9. Before attacking parts a and b, first draw a Venn diagram and attempt to fill in as much data as possible.

Here is the basic diagram which needs to be filled in. M stands for the set of people who own microwave ovens, A stands for the set of people who own answering machines, and V stands for the set of people who own VCR's.

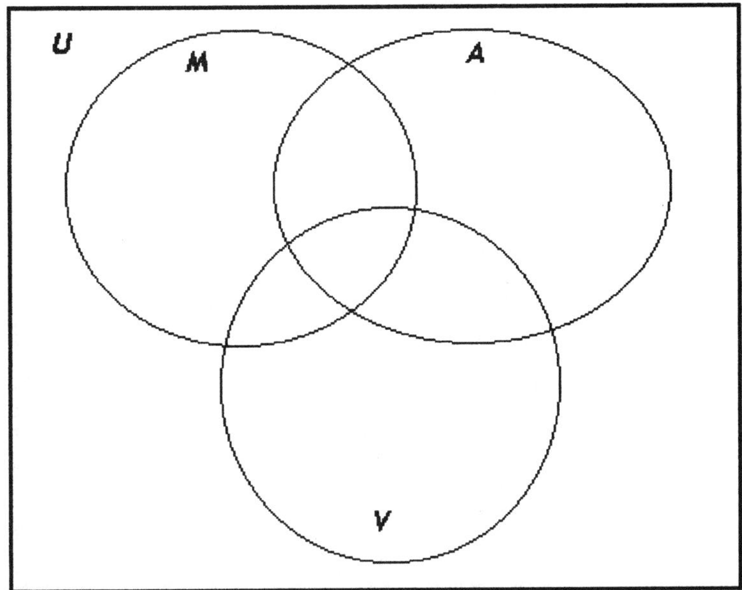

Now we fill in the blanks. We can immediately enter that 69 had all three, 64 had none, 57 had answering machines but no microwaves or VCR's, and 104 had microwaves and VCR's but no answering machines. *Caution:* It will take some computation before we can enter that 98 had answering machines and VCR's, because that figure covers those with or without microwaves.

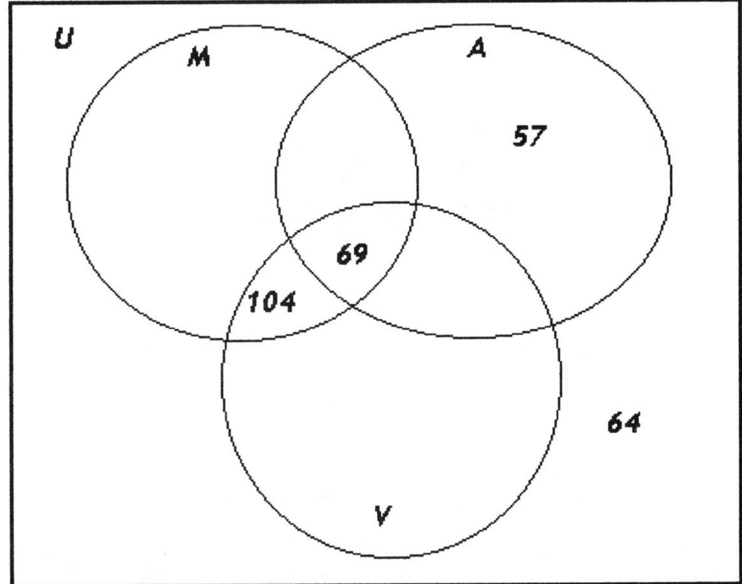

Now, 98 people had answering machines and VCR's, so 98 − 69 = 29 had answering machines and VCR's but no microwaves.

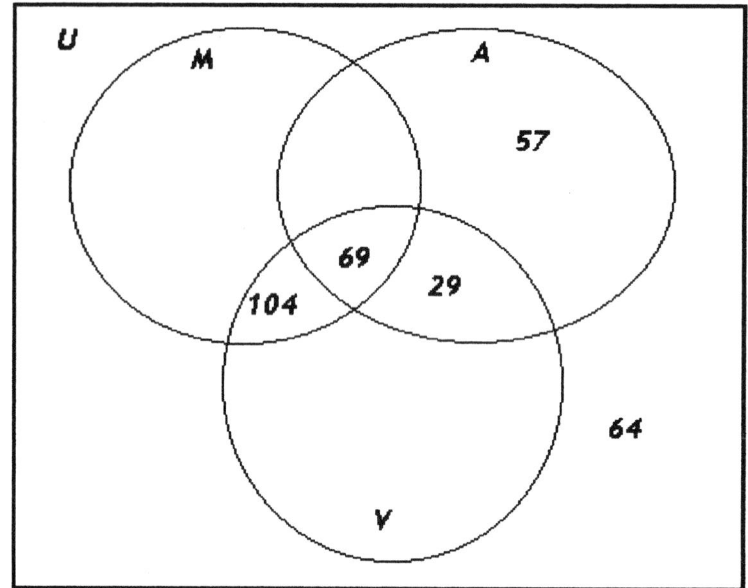

The number of people who owned only answering machines and microwaves was $232-(57+29+69)=77$.

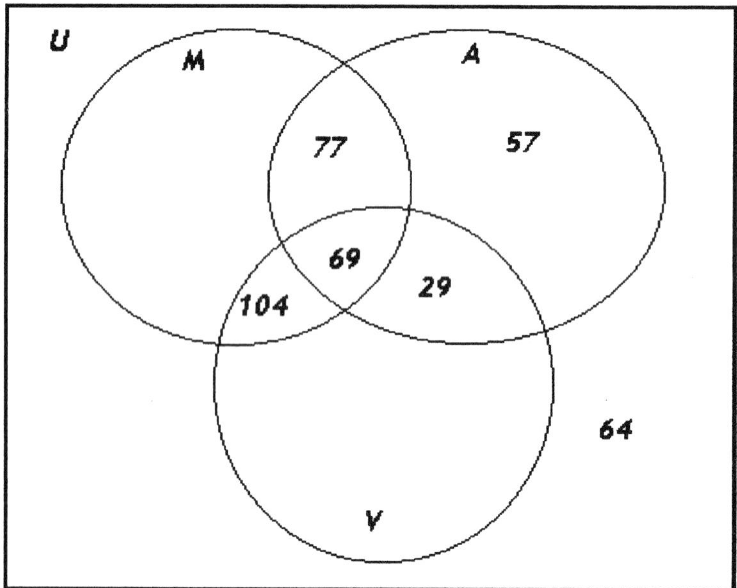

The number of people who owned only microwaves was $313-(104+69+77)=63$, and the number of people who owned only VCR's was $269-(104+69+29)=67$.

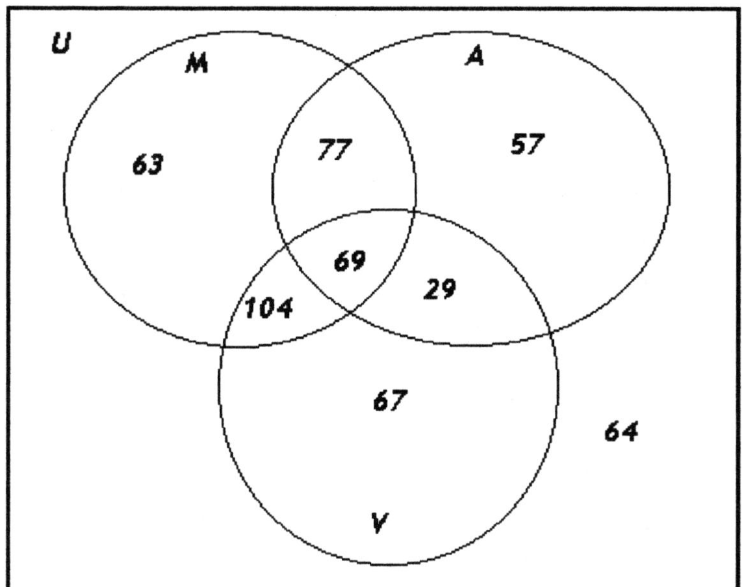

Now all of the blanks are filled in and we can tackle parts a and b with confidence.

9a. First we need the number of people surveyed. This is found by adding all of the numbers in the final Venn diagram above:

$$63 + 77 + 57 + 104 + 69 + 29 + 67 + 64 = 530$$

The number of people who owned only a microwave was 63, so the percentage we seek is $\frac{63}{530} \cdot 100 \approx 11.89$.

9b. $\frac{67}{530} \cdot 100 \approx 12.64\%$

13. First, use the fact the the number of people who used black and white or color equals the number of people who used black and white plus the number of people who used color minus the number of people who used black and white and color, i.e.,

$$n(B\&W \cup C) = n(B\&W) + n(C) - n(B\&W \cap C)$$
$$101 = 77 + 65 - 41$$

Thus we obtain the following Venn diagram:

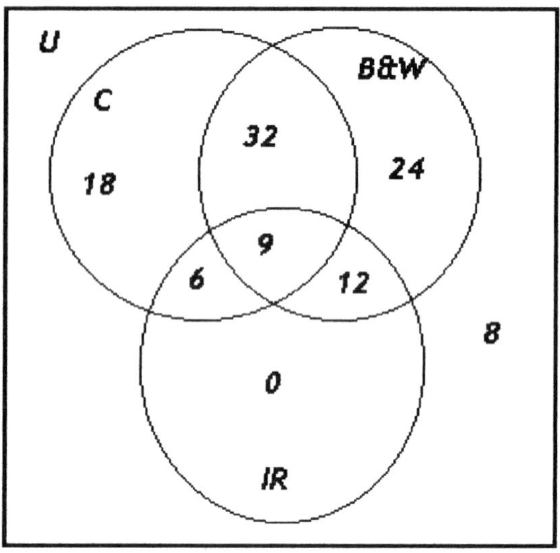

13a. 0%

13b. $\frac{32+6+12+9}{109} \cdot 100 \approx 54.13\%$

17. Let F be the set of fish owners, B be the set of bird owners, C be the set of cat owners, and D be the set of dog owners. We can establish the following Venn diagram immediately without computations:

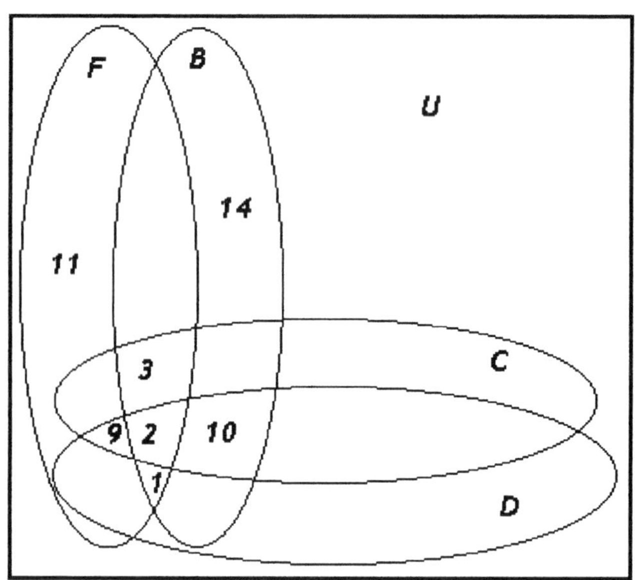

Since 10 people own a fish and a bird, the number of people who own *only* a fish and a bird is $10-(3+2+1)=4$:

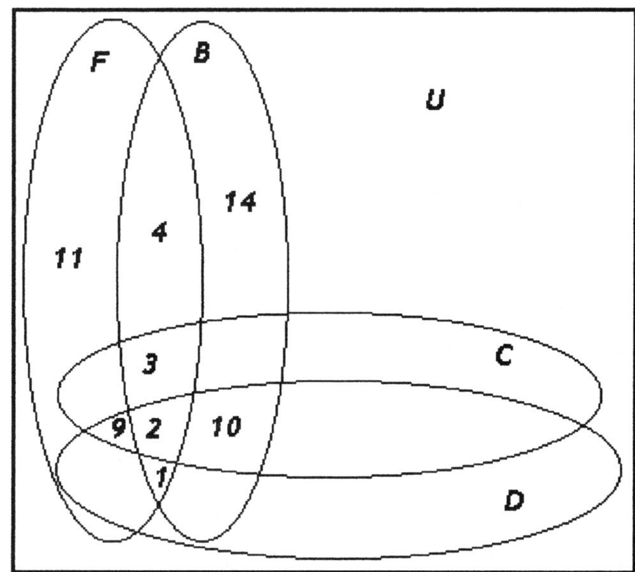

Since 21 own a fish and a cat, the number who own *only* a fish and a cat is $21-(3+2+9)=7$:

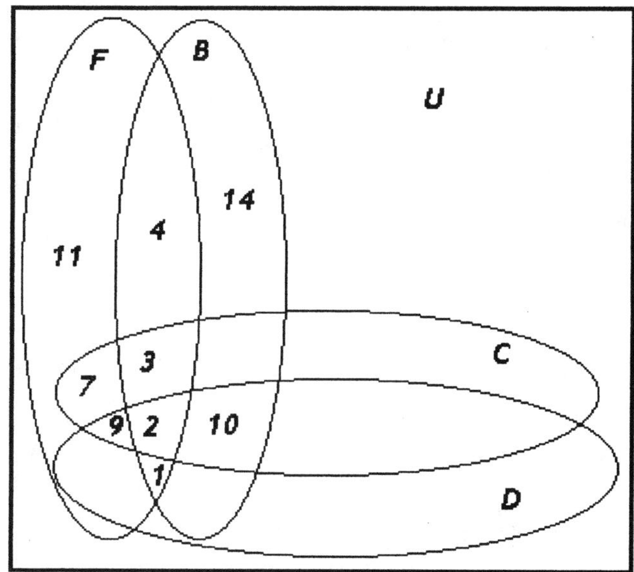

Since 26 own a bird and a dog, then $26-(10+2+1)=13$ own only a bird and a dog:

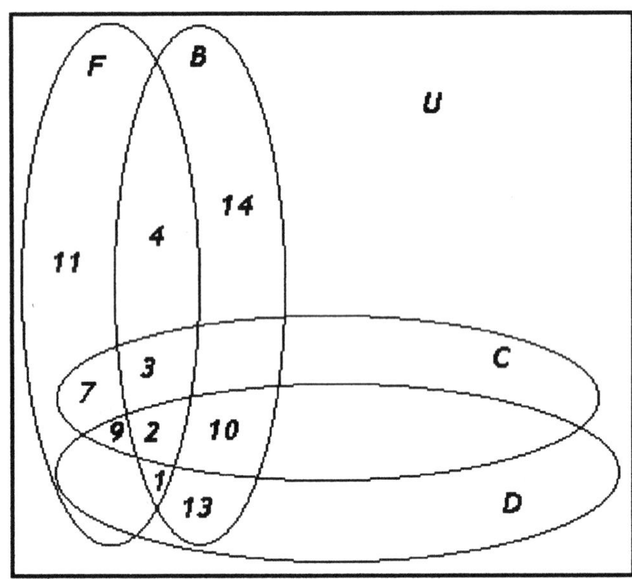

Since 27 own a cat and a dog, then $27-(9+2+10)=6$ own only a cat and a dog:

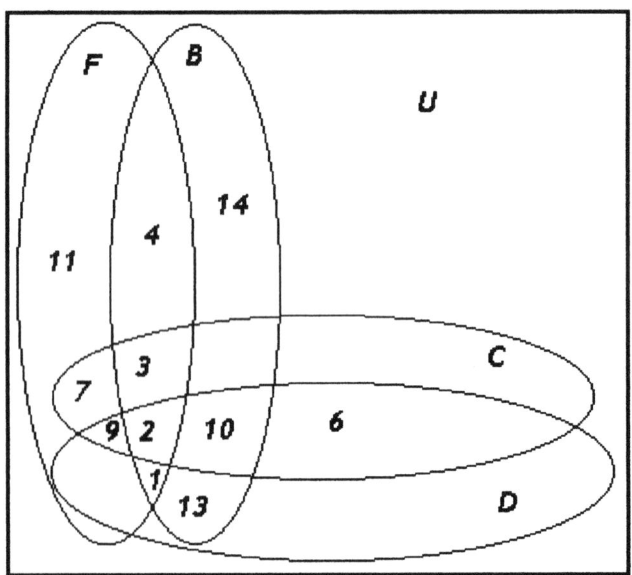

Since 49 own fish, then $49-(11+4+7+3+9+2+1)=12$ own only fish and a dog:

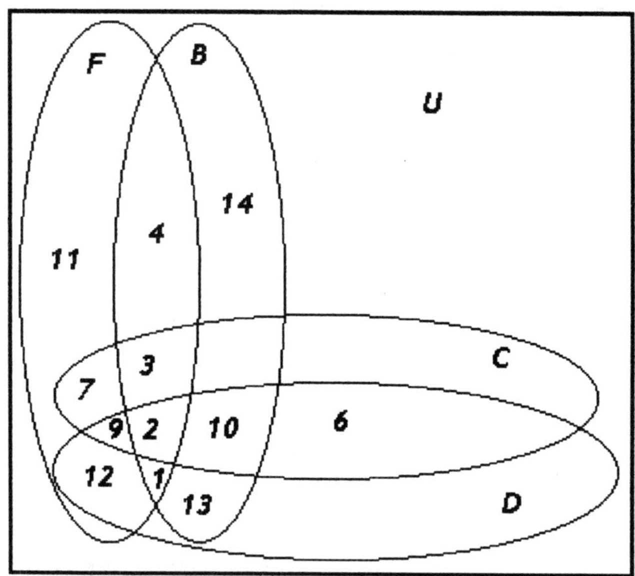

Since 55 own a bird, then $55 - (14 + 4 + 3 + 2 + 1 + 13 + 10) = 8$ own only a bird and a cat:

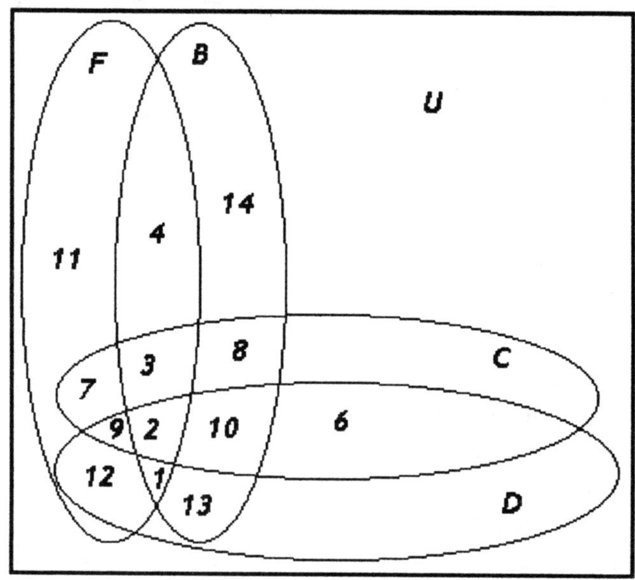

Since 50 own a cat, then $50 - (8 + 3 + 7 + 9 + 2 + 10 + 6) = 5$ own only a cat:

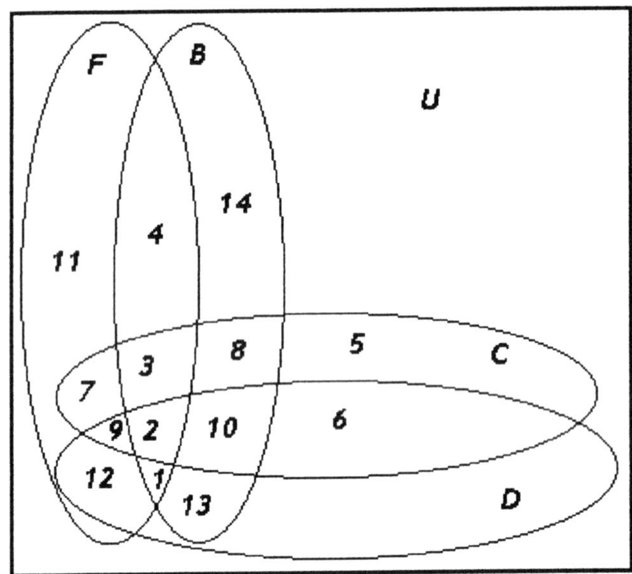

Since 68 own a dog, then $68 - (13+1+12+9+2+10+6) = 15$ own only a dog:

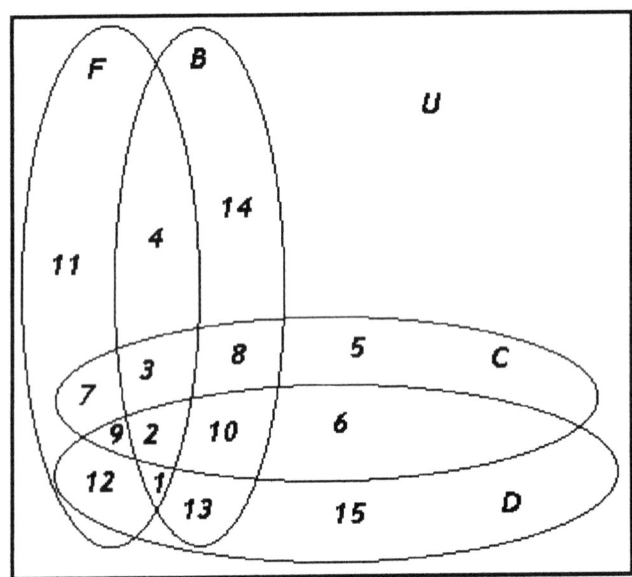

Summing all of the numbers gives us $120 = n(F \cup B \cup C \cup D)$. Thus $136 - 120 = 16$ pet owners have no fish, no birds, no cats, and no dogs.

21.

$$\begin{aligned}(A \cap B')' &= A' \cup (B')' \\ &= A' \cup B \\ &= \{1,3,6,7,8\} \cup \{1,2,7,8,9\} \\ &= \{1,2,3,6,7,8,9\}\end{aligned}$$

25.

$(A \cup B)'$

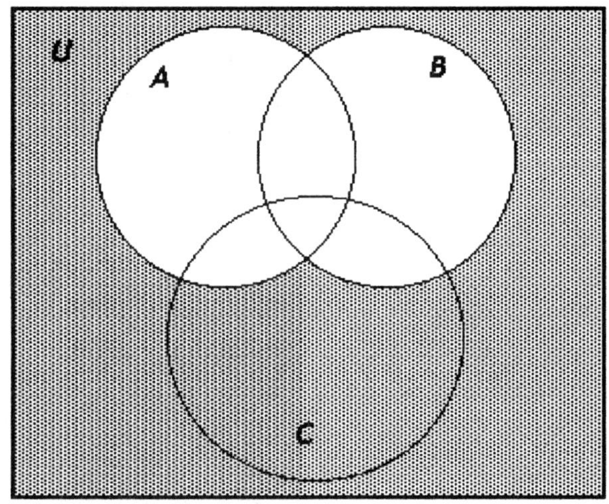

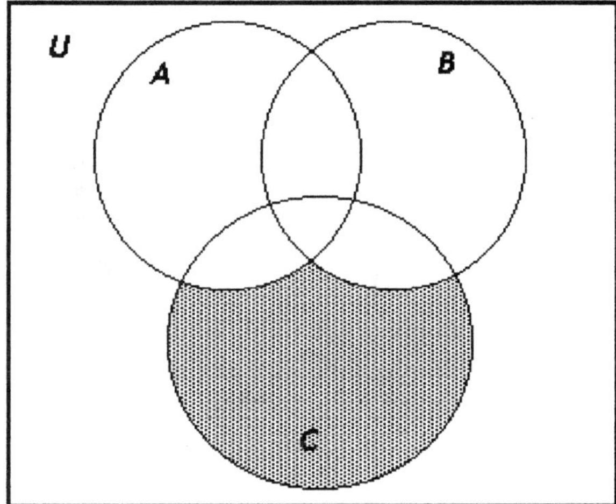

$(A \cup B)' \cap C$

29.

$(A \cap B)'$

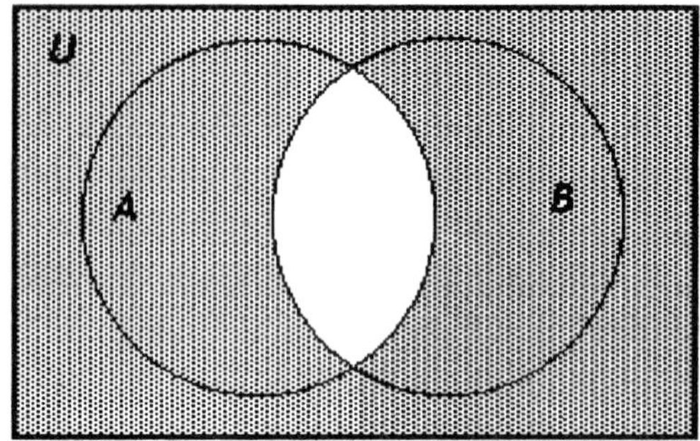

A'

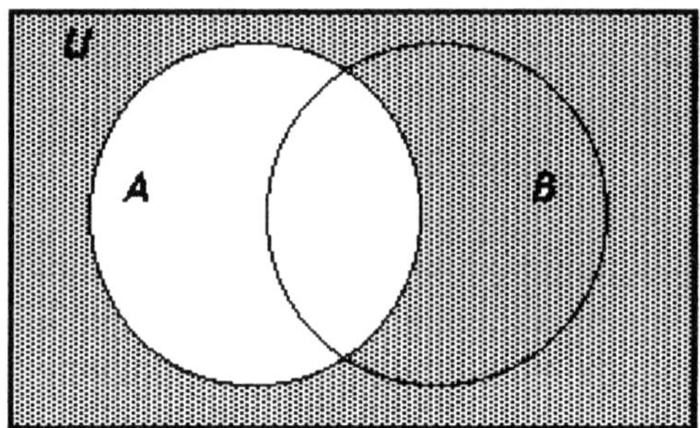

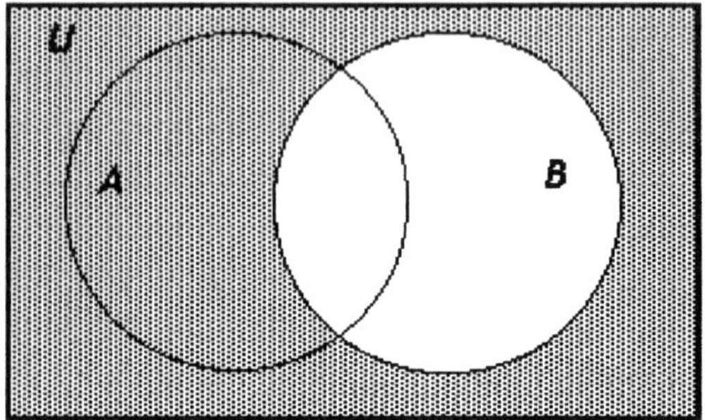

B'

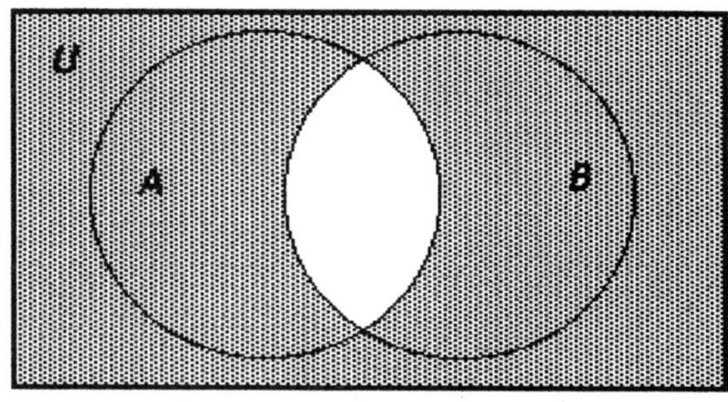

$A' \cup B'$

33. If all members attend, it can't be held at Betty's or Delores' houses. It also can't be held on Tuesday or Friday. These conditions eliminate possibilities a, b, c, and e. Thus d is the only possible meeting time and place out of the ones listed.

37. Since the meeting isn't at Frank's house, the only other possible meeting places are Betty's house and Delores' house. Since Frank is attending, then Angela and Carmine are also attending, but Angela can't attend meetings at Delores' house, so the meeting must be at Betty's house. Selection b must be true. (This means that Betty is attending, so there are at least 3 women, which rules out selection a.)

Section 5.3
Introduction to Combinatorics

1a. $2^3 = 8$

1b.

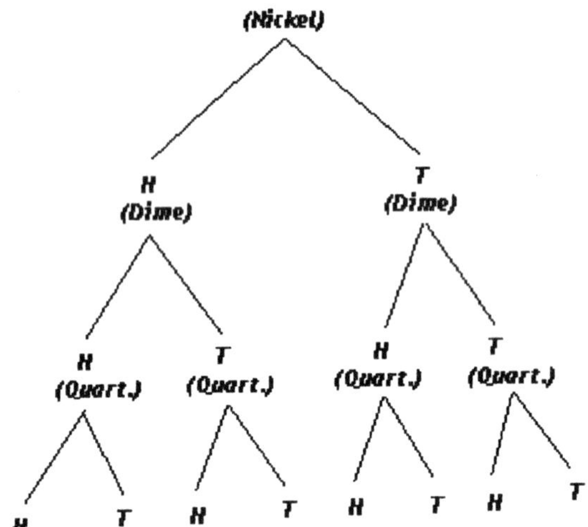

3a. $2 \cdot 3 \cdot 2 = 12$

3b.

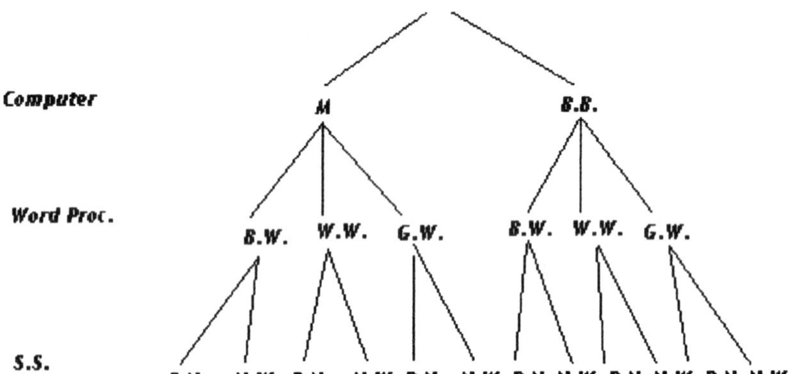

5. $3 \cdot 4 \cdot 2 = 24$

9. $14 \cdot 7 \cdot 9 \cdot 3 = 2646$

13. $10^9 = 1,000,000,000$ (This assumes that numbers such as 000-00-0000 are permissible.)

17. There are 8 possibilities for the first and third digits, and 2 possibilities for the second digit; thus there are $8 \cdot 2 \cdot 8 = 128$ possible area codes.

21. $4! = 4 \cdot 3 \cdot 2 \cdot 1 = 24$

25. $20! = 20 \cdot 19 \cdot 18 \cdot \cdots \cdot 2 \cdot 1 \approx 2.4329 \times 10^{18}$

29a. $\frac{6!}{4!} = \frac{6 \cdot 5 \cdot 4!}{4!} = 6 \cdot 5 = 30$

29b. $\frac{6!}{2!} = \frac{6 \cdot 5 \cdot 4 \cdot 3 \cdot 2!}{2!} = 360$

33. $\frac{8!}{4! \cdot 4!} = \frac{8 \cdot 7 \cdot 6 \cdot 5 \cdot 4!}{4! \cdot 4!} = \frac{1680}{4!} = 70$

37. $\frac{16!}{(16-14)!} = \frac{16!}{2!} \approx 1.0461 \times 10^{13}$

41. $\frac{7!}{(7-3)! \cdot 3!} = \frac{7!}{4! \cdot 3!} = 35$

45. If a procedure can be broken into n successive stages, with a_1 outcomes in the first stage, a_2 outcomes in the second stage, ..., a_n outcomes in the nth stage, then the total number of composite outcomes is $a_1 \cdot a_2 \cdot \cdots \cdot a_n$. (We assume that the composite outcomes are distinct.)

49. The gray car is in space 2 and the green car is in space 3. That leaves the black, white, and yellow cars. The black car can't be in space 1 because it must be next to the yellow car. The white car also can't be in space 1 because the gray car can't be next to the white car. That leaves the yellow car for space 1, but that isn't permissible because the yellow car must be next to the black car. Thus none of the cars can be parked in space 1.

Section 5.4
Permutations and Combinations

1a. ${}_7P_3 = \frac{7!}{(7-3)!} = \frac{7\cdot6\cdot5\cdot4!}{4!} = 7\cdot6\cdot5 = 210$

1b. ${}_7C_3 = \frac{7!}{(7-3)!\cdot3!} = \frac{7\cdot6\cdot5\cdot4!}{4!3!} = \frac{7\cdot6\cdot5}{3\cdot2\cdot1} = 35$

5a. ${}_{14}P_1 = \frac{14!}{(14-1)!} = \frac{14\cdot13!}{13!} = 14$

5b. ${}_{14}C_1 = \frac{14!}{(14-1)!\cdot1!} = \frac{14!}{(14-1)!} = 14$

9a. ${}_xP_{x-1} = \frac{x!}{[x-(x-1)]!} = \frac{x!}{1!} = x!$

9b. ${}_xC_{x-1} = \frac{x!}{[x-(x-1)]!\cdot(x-1)!} = \frac{x!}{1!\cdot(x-1)!} = \frac{x!}{(x-1)!} = \frac{x(x-1)!}{(x-1)!} = x$

13a. ${}_3P_2 = \frac{3!}{(3-2)!} = \frac{6}{1} = 6$

13b.

$$(a,b),\\(a,c),\\(b,c),\\(b,a),\\(c,a),\\(c,b)$$

17a. There are 12 choices for the first presenter, 11 choices for the second presenter, and so on. Thus there are $12! = 479,001,600$ different presentation orders possible.

17b. If their names are put into alphabetical order, where the "a's" go first, there is only one presentation order possible, unless, of course, more than one person has exactly the same name.

21. This is equivalent to forming all possible combinations (without regard to order) of 14 objects taken 2 at a time, or ${}_{14}C_2 = 91$. For example, if we have only 3 teams, say a, b, and c, the possible pairings for games are a, b; a, c; b, c. That is, we took the number of combinations of 3 objects taken 2 at a time, or ${}_3C_2 = 3$.

25a. Two women out of the 8 and 2 men out of the 6 are to be selected without regard to order; i.e., $({}_8C_2)({}_6C_2) = (28)(15) = 420$.

25b. ${}_{14}C_4 = 1001$

25c. In order for the 4-person committee to have a majority of women, it must consist of 3 women and 1 man or 4 women and no men. In the former case we have a total of $({}_8C_3)({}_6C_1)$ and in the latter case we have a total of $({}_8C_4)({}_6C_0)$. Adding these gives the grand total:
$({}_8C_3)({}_6C_1) + ({}_8C_4)({}_6C_0) = (56)(6) + (70)(1) = 406.$

Use the following table for Exercises 27 - 31.

BLACK		RED	
♠ SPADES	♣ CLUBS	♥ HEARTS	♦ DIAMONDS
Ace	Ace	Ace	Ace
King	King	King	King
Queen	Queen	Queen	Queen
Jack	Jack	Jack	Jack
10	10	10	10
9	9	9	9
8	8	8	8
7	7	7	7
6	6	6	6
5	5	5	5
4	4	4	4
3	3	3	3
2	2	2	2

29a. $({}_4C_3)({}_{48}C_2) = 4512$ (Note that the second factor involves choosing 2 cards out of *48* and not 49. Since we want *exactly* 3 Aces, after those are chosen, 1 Ace remains, but this cannot be part of the next selection.)

29b. The number of hands with exactly 3 Aces equals the number of hands with exactly 3 Kings (or any other denomination), so there are $13 \cdot ({}_4C_3)({}_{48}C_2) = 58,656$ ways.

However, the poker hand *three of a kind* means a hand with 3 cards of the same denomination, such as 3 Kings, 3 8's, and so on, without regard to suit. The other two cards must be of different denominations. For example, Q, Q, Q, 7, 10, or J, J, J, 4, 8 are two such hands. However, Q, Q, Q, 5, 5 is not such a hand; it is a *full house*. We break this procedure into component parts.

1) Choose the denomination for the three of a kind from the 13 available denominations. This can be done in ${}_{13}C_1 = 13$ ways.

2) Choose 3 suits for the three of a kind (such as heart, spade, and club). This can be done in ${}_4C_3 = 4$ ways. This completes the selection for the three of a kind; now 2 cards remain to be chosen.

3) Choose the 2 denominations for the other two cards in the hand. There are 12 denominations left from which to choose. This can be done in ${}_{12}C_2 = 66$ ways.

4) Choose the suit for each denomination chosen in 3) above. For example, if 2 and K were chosen in part 3), one could choose hearts for both, or spades and clubs. There are $4^2 = 16$ ways to do this.

By the Fundamental Principle of Counting, there are $13 \cdot 4 \cdot 66 \cdot 16 = 54,912$ poker hands with three of a kind.

One might wonder, after step 2), why not just choose 2 cards out of the 48 remaining as in Exercise 29a $({}_{48}C_2)$and then be done with it? If this is done, one might choose two cards of the same denomination, which will result in a full house.

33. ${}_{53}C_6 = 22,957,480$

37. A 5/36 lottery is easier to win because fewer choices means a greater probability of winning.

41a. Row 4.

41b. Row n.

41c. No.

41d. Yes.

41e. It is the $(r+1)$st number in row n. The first number in each row corresponds to $r = 0$.

45. On the first Friday B plays E, and on the third Friday B plays C; on the second Friday A plays F. Thus on the second Friday the following teams have either already played (or will play later) with B, or are occupied in other games: E, C, A, and F. That leaves team D to play with team B on the second Friday.

Chapter 5
Review Exercises

1. (See Sec. 5.1 for information on John Venn and Sec. 5.2 for information on Augustus De Morgan.)

2a. $A' = \{1, 3, 5, 7, 9\} = B$

2b. $B' = \{0, 2, 4, 6, 8\} = A$

2c. $A \cup B = \{0, 1, 2, 3, 4, 5, 6, 7, 8, 9\} = U$

2d. $A \cap B = \{\ \} = \emptyset$

3a. $A \cup B =$ {Maria, Nobuko, Leroy, Mickey, Kelly, Rachel, Deanna}

3b. $A \cap B =$ {Leroy, Mickey}

4. Proper: {Dallas, Chicago}, {Dallas, Tampa}, {Chicago, Tampa}, {Dallas}, {Chicago}, {Tampa}, $\emptyset$.
Improper: C

5a.

$$\begin{aligned} n(A \cap B) &= n(A) + n(B) - n(A \cup B) \\ &= 32 + 26 - 40 \\ &= 18 \end{aligned}$$

5b.

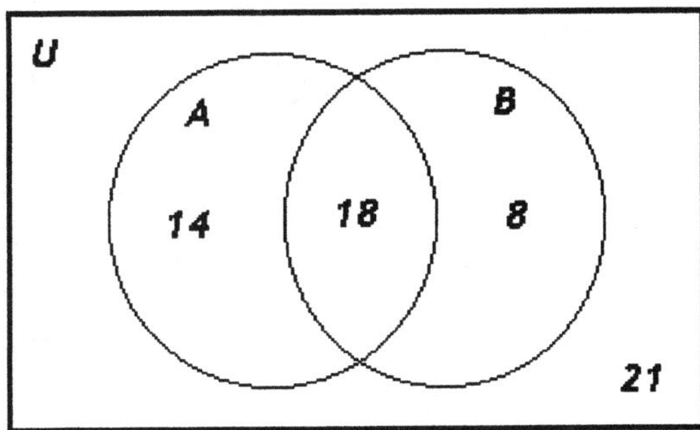

6.

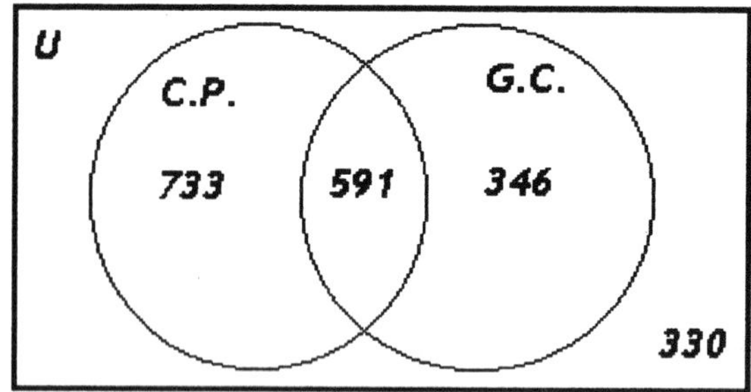

6a. $733 + 591 + 346 = 1670$

6b. 733

6c. 346

6d. 330

7.

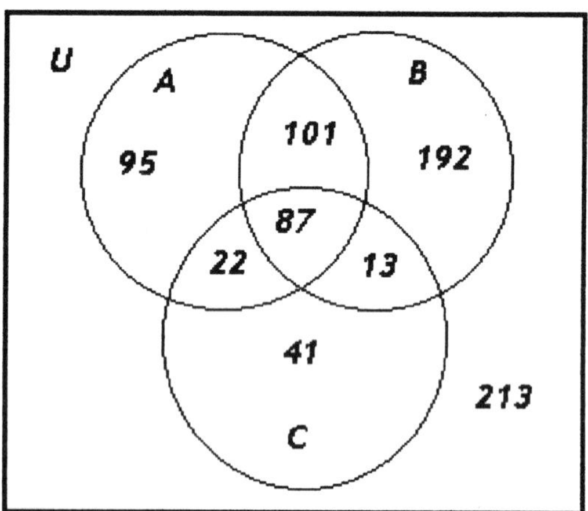

$\frac{101+22+13+87}{764} \cdot 100 \approx 29.19\%$

8a.

$$\begin{aligned}(A' \cup B)' &= (A')' \cap B' \\ &= A \cap B' \\ &= \{b, d, f, g\} \cap \{b, e, f, h\} \\ &= \{b, f\}\end{aligned}$$

8b.

$$\begin{aligned}(A \cap B')' &= A' \cup (B')' \\ &= A' \cup B \\ &= \{a, c, e, h, i\} \cup \{a, c, d, g, i\} \\ &= \{a, c, d, e, g, h, i\}\end{aligned}$$

9a. $2 \cdot 3 \cdot 2 = 12$

9b.

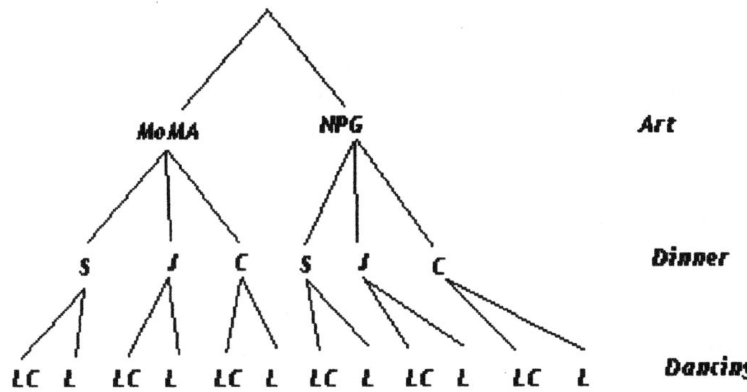

10. $5 \cdot 3 \cdot 3 \cdot 2 \cdot 2 = 180$

11. $9 \cdot 10^4 \cdot 4 \cdot 3 = 1,080,000$

12a. $10! = 3,628,800$

12b. $0! = 1$

12c. $\frac{82!}{79!} = \frac{82 \cdot 81 \cdot 80 \cdot 79!}{79!} = 531,360$

12d. $\frac{27!}{20!7!} = \frac{27 \cdot 26 \cdot \cdots \cdot 20!}{20!7!} = \frac{4,475,671,200}{5040} = 888,030$

13a. ${}_{11}C_3 = 165$

13b. ${}_{11}P_3 = 990$

14a. ${}_{15}P_4 = 32,760$

14b. ${}_{15}C_4 = 1365$

14c. ${}_{15}P_{11} = 54,486,432,000$

15a. $({}_{10}C_1)({}_{12}C_2) = 660$

15b. ${}_{22}C_3 = 1540$

15c. $({}_{10}C_1)({}_{12}C_2) + ({}_{10}C_0)({}_{12}C_3) = 880$

16. ${}_{10}C_2 = 45$

17. ${}_{10}P_3 = 720$

18. ${}_{52}C_7 = 133,784,560$

19. ${}_{54}C_7 = 177,100,560$

20. ${}_{42}C_6 = 5,245,786$

21. Order

22a. 4th entry, row 7

22b. 5th entry, row 7

22c. ${}_7C_3 = \frac{7!}{(7-3)!3!} = \frac{7!}{4!3!} = \frac{7!}{(7-4)!4!} = {}_7C_4$

22d. Row n, entry $(r+1)$

23a. ${}_3C_1 = 3$

23b. ${}_3C_2 = 3$

23c. ${}_3C_3 = 1$

23d. ${}_3C_0 = 1$

23e. 8

23f. The number of subsets of $S = 2^{n(S)}$.

Chapter 6 Probability

Section 6.1 Introduction to Probability

1. The experiment is the act of picking a jelly bean.

5. Since the events are mutually exclusive, $p(\text{red or yellow}) = p(\text{red}) + p(\text{yellow}) = \frac{n(\text{red})}{n(S)} + \frac{n(\text{yellow})}{n(S)} = \frac{n(\text{red})+n(\text{yellow})}{n(S)} = \frac{8+10}{35} = \frac{18}{35} \approx 0.51.$

9. $p(\text{white}) = \frac{n(\text{white})}{n(S)} = \frac{0}{35} = 0$

13. 18:17

BLACK		RED	
♠ SPADES	♣ CLUBS	♥ HEARTS	♦ DIAMONDS
Ace	Ace	Ace	Ace
King	King	King	King
Queen	Queen	Queen	Queen
Jack	Jack	Jack	Jack
10	10	10	10
9	9	9	9
8	8	8	8
7	7	7	7
6	6	6	6
5	5	5	5
4	4	4	4
3	3	3	3
2	2	2	2

17a. $4 \times 3 = 12$

17b. $\frac{12}{52} = \frac{3}{13}$

21. Drawing the card.

25a. $\frac{4}{52} = \frac{1}{13}$

25b. 1:12

29a. "Count an ace as high" means to assign it a value of 11. An ace can be worth either 1 or 11 points; face cards are worth 10 points.
$\frac{12}{52} = \frac{3}{13}$

29b. 3:10

33a. $\frac{12}{52} = \frac{3}{13}$

33b. 3:10

37. a:(b – a)

41a. $\frac{26}{52} = \frac{1}{2}$

41b. 1:1

45a. $S = \{(b,b,b),(b,b,g),(b,g,b),(g,b,b),(b,g,g),(g,b,g),(g,g,b),(g,g,g)\}$

45b. $E = \{(b,g,g),(g,b,g),(g,g,b)\}$

45c. $F = \{(b,g,g),(g,b,g),(g,g,b),(g,g,g)\}$

45d. $G = \{(g,g,g)\}$

45e. $p(E) = \frac{3}{8}$

45f. $p(F) = \frac{4}{8} = \frac{1}{2}$

45g. $p(G) = \frac{1}{8}$

45h. $o(E) = 3:5$

45i. $o(F) = 1:1$

45j. $o(G) = 1:7$

49. (Cf. Exercise 45.) Having three children of different sexes is more likely.

61. Gombauld was a French gentleman who consulted with Blaise Pascal, the famous mathematician, philosopher, and theologian, about a problem involving the throwing of dice, thus initiating the first well-known mathematical studies in the theory of probability.

65. Calculating the relative frequency would involve actual trials.

Probability Distributions

1a.

Outcome	Probability
spin a 1	$\frac{36}{360} = \frac{1}{10}$
spin a 2	$\frac{72}{360} = \frac{1}{5}$
spin a 3	$\frac{108}{360} = \frac{3}{10}$
spin a 4	$\frac{144}{360} = \frac{2}{5}$

1b.

$$\begin{aligned} p(\text{not spinning a 1}) &= p(\text{spinning a 2 or 3 or 4}) \\ &= \frac{1}{5} + \frac{3}{10} + \frac{2}{5} \\ &= \frac{9}{10} \end{aligned}$$

5a.

	r	*r*
r	rr	rr
w	rw	rw

Outcome	Probability
rr	$\frac{1}{2}$
rw	$\frac{1}{2}$
ww	0

5b. $\frac{1}{2}$

5c. $\frac{1}{2}$

5d. 0

9a. Let s represent the disease-free gene and S represent the sickle-cell gene.

	S	s
S	SS	Ss
s	Ss	ss

Outcome	Probability
ss	$\frac{1}{4}$
Ss	$\frac{1}{2}$
SS	$\frac{1}{4}$

9b. $\frac{1}{4}$

9c. $\frac{1}{2}$

9d. $\frac{1}{4}$

13a. Let h represent the disease-free gene and H represent the dominant Huntington's disease gene.

	H	h
h	Hh	hh
h	Hh	hh

Outcome	*Probability*
hh	$\frac{1}{2}$
Hh	$\frac{1}{2}$
HH	0

13b. $\frac{1}{2}$

13c. $\frac{1}{2}$

13d. $\frac{1}{2}$

Section 6.3
Basic Rules of Probability

1. The events are not mutually exclusive; it is possible to be both a doctor and a woman.

5. The events are not mutually exclusive; it is possible to have both brown and gray hair. This, however, involves playing semantic games. When one says, "I have brown hair," does one mean that his or her hair is naturally brown or artificially-colored gray hair? (A mix of the two colors is also possible, but then it is neither brown nor gray.)

9. The events are mutually exclusive; four is not an odd number.

BLACK		RED	
♠ SPADES	♣ CLUBS	♥ HEARTS	♦ DIAMONDS
Ace	Ace	Ace	Ace
King	King	King	King
Queen	Queen	Queen	Queen
Jack	Jack	Jack	Jack
10	10	10	10
9	9	9	9
8	8	8	8
7	7	7	7
6	6	6	6
5	5	5	5
4	4	4	4
3	3	3	3
2	2	2	2

13a. $\frac{1}{52}$

13b. $\frac{4+13-1}{52} = \frac{4}{13}$

13c. $1 - \frac{1}{52} = \frac{51}{52}$

17a. $\frac{36}{52} = \frac{9}{13}$

17b. $\frac{32}{52} = \frac{8}{13}$

17c. $\frac{16}{52} = \frac{4}{13}$

17d. $\frac{36+32-16}{52} = 1$

21. $1 - \frac{12}{52} = \frac{10}{13}$

25. $1 - \frac{16}{52} = \frac{9}{13}$

29. If $p(E) = \frac{2}{7} = \frac{2}{2+5}$, then $o(E) = 2:5$ and $o(E') = 5:2$.

33. p(king) $= \frac{4}{52} = \frac{1}{13}$; o(not a king) $= (13-1):1 = 12:1$

37. p(four or below) $= \frac{12}{52} = \frac{3}{13}$; o(above a four) $= 10:3$

41a. $\frac{151}{700} \approx 0.22$

41b. $\frac{71}{175} + \frac{88}{175} - \frac{151}{700} = \frac{97}{140} \approx 0.69$

45a. $\frac{183+177}{1000} = \frac{9}{25} = 0.36$

45b. $1 - \frac{9}{25} = \frac{16}{25} = 0.64$

Note that when rolling a pair of dice, the sample space consists of $6 \times 6 = 36$ ordered pairs:
$S = \{(1,1),(1,2),(1,3),(1,4),(1,5),(1,6),(2,1),(2,2),(2,3),(2,4),(2,5),(2,6),(3,1),\ldots,(3,6),(4,1),\ldots,(4,6),(5,1),\ldots,(5,6),(6,1),\ldots,(6,6)\}$.

49a. $\frac{6+2}{36} = \frac{2}{9}$

49b. (Note that doubles, e.g., $(2,2)$ or $(5,5)$, never add up to 7 or 11.)
$\frac{2}{9} + \frac{6}{36} = \frac{7}{18}$

53a. $\frac{6}{36} = \frac{1}{6}$

53b. $\frac{1+3+5+5+3+1}{36} + \frac{1}{6} - \frac{1}{6} = \frac{1}{2}$ (All doubles sum to even numbers)

57. Let A be the event "defective bulb" and B the event "defective battery."

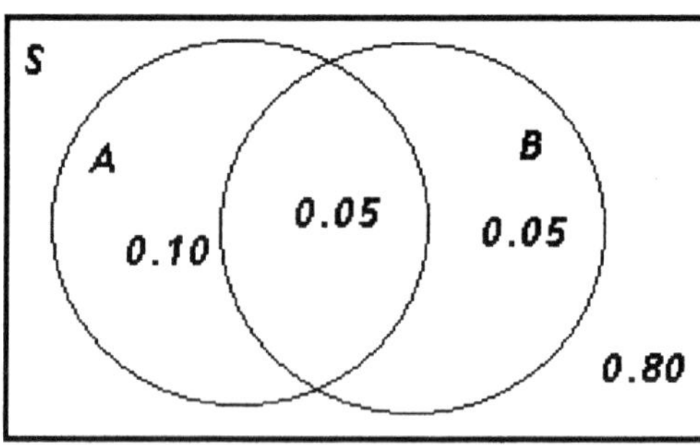

57a. $p(A \cup B) = 0.20$

57b. $p(A' \cup B') = p((A \cap B)') = 1 - p(A \cap B) = 1 - 0.05 = 0.95$

57c. $p(A' \cap B') = p((A \cup B)') = 0.8$

61.

$$\begin{aligned} n(E \cup F) &= n(E) + n(F) - n(E \cap F) \\ \frac{n(E \cup F)}{n(S)} &= \frac{n(E) + n(F) - n(E \cap F)}{n(S)} \\ &= \frac{n(E)}{n(S)} + \frac{n(F)}{n(S)} - \frac{n(E \cap F)}{n(S)} \\ p(E \cup F) &= p(E) + p(F) - n(E \cap F) \end{aligned}$$

65. The "empty" event; an impossible event.

Combinatorics and Probability

1. $1 - \frac{_{365}P_{30}}{365^{30}} \approx 0.706$

5.

$$\frac{_6C_5 \cdot {_{38}C_1}}{_{44}C_6} \approx 0.0000323.$$

9a. $\frac{1}{_{39}C_5} = \frac{1}{575{,}757} \approx 0.0000017$

9b.

$$\frac{_5C_4 \cdot {_{34}C_1}}{_{39}C_5} \approx 0.000295263.$$

13. In 8-spot Keno, for example, there are $_{80}C_8$ ways of marking an 8-spot card. With 8 specific numbers marked and 20 numbers randomly drawn, there are $_{20}C_8$ (out of the $_{80}C_8$) ways for all of the 8 specifically marked numbers to be present among the 20 numbers drawn. Hence the probability that all 8 marked numbers are drawn is

$$\frac{_{20}C_8}{_{80}C_8} \approx 0.0000043.$$

Outcome	*Probability*
8 winning spots	$\frac{_{20}C_8}{_{80}C_8} \approx 0.0000043$
7 winning spots	$\frac{_{20}C_7 \cdot {_{60}C_1}}{_{80}C_8} \approx 0.000160455$
6 winning spots	$\frac{_{20}C_6 \cdot {_{60}C_2}}{_{80}C_8} \approx 0.002366714$
5 winning spots	$\frac{_{20}C_5 \cdot {_{60}C_3}}{_{80}C_8} \approx 0.0183025856$
4 winning spots	$\frac{_{20}C_4 \cdot {_{60}C_4}}{_{80}C_8} \approx 0.0815037015$
$<$ 4 winning spots	1 – (sum of above) $\approx$ 0.897662

Optional: For Exercises 17-23 refer to the following tree diagram, where P indicates a burrito with peppers and P' indicates a burrito without peppers:

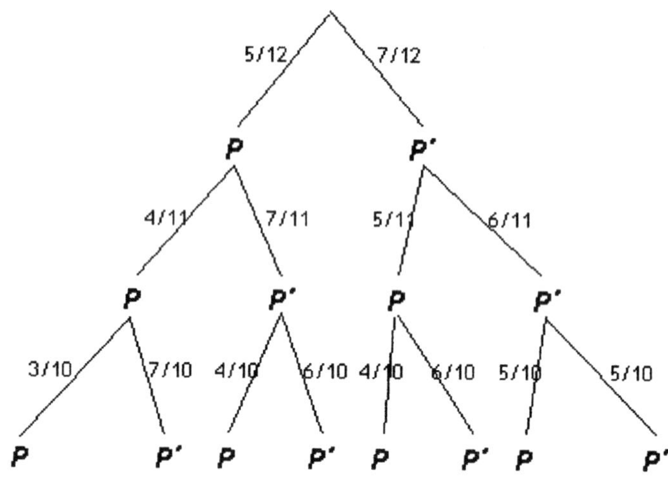

17.

$$\frac{{}_5C_3 \cdot {}_7C_0}{{}_{12}C_3} \approx 0.045$$

Here is an alternate way of solving it. Before choosing a burrito, there are 5 out of 12 burritos with peppers. After choosing a burrito with peppers, there will be 4 out of 11 burritos with peppers. Thus the probability of picking 3 burritos with peppers is $\left(\frac{5}{12}\right)\left(\frac{4}{11}\right)\left(\frac{3}{10}\right) = \frac{1}{22}$. (See the tree diagram above.)

21.

$$p(\text{no hot peppers}) + p(\text{1 hot pepper}) =$$
$$\frac{{}_5C_0 \cdot {}_7C_3}{{}_{12}C_3} + \frac{{}_5C_1 \cdot {}_7C_2}{{}_{12}C_3} \approx 0.64.$$

Here is an alternate way of solving it. (This includes none with peppers.) (See the tree diagram above.)
$\left(\frac{5}{12}\right)\left(\frac{7}{11}\right)\left(\frac{6}{10}\right) + \left(\frac{7}{12}\right)\left(\frac{5}{11}\right)\left(\frac{6}{10}\right) + \left(\frac{7}{12}\right)\left(\frac{6}{11}\right)\left(\frac{5}{10}\right) + \left(\frac{7}{12}\right)\left(\frac{6}{11}\right)\left(\frac{5}{10}\right) = \frac{7}{11}$

25a. (Make a tree diagram like that above.)

$$\frac{{}_{140}C_0 \cdot {}_{60}C_2}{{}_{200}C_2} \approx 0.08894$$

25b.

$$\frac{{}_{140}C_1 \cdot {}_{60}C_1}{{}_{200}C_2} \approx 0.42211$$

25c.

$$\frac{{}_{140}C_2 \cdot {}_{60}C_0}{{}_{200}C_2} \approx 0.48894$$

Section 6.5
Probability Distributions and Expected Value

1. $0(0.15) + 1(0.35) + 2(0.25) + 3(0.15) + 4(0.05) + 5(0.05) = 1.75$

5.

$$\begin{aligned}(35)\left(\frac{1}{38}\right) + (-1)\left(\frac{37}{38}\right) &= \frac{35}{1}\cdot\frac{1}{38} + \frac{-1}{1}\cdot\frac{37}{38}\\ &= \frac{35\cdot 1}{1\cdot 38} + \frac{-1\cdot 37}{1\cdot 38}\\ &= \frac{35}{38} + \frac{37(-1)}{38}\\ &= \frac{35+37(-1)}{38}\end{aligned}$$

9. Let x be the probability of success. (Note that if x is the probability of success, then the complementary probability $1 - x$ is the probability of failure.)

Outcome	**Value**	**Probability**
Success	x	0.50
Failure	$1 - x$	-0.60

$$\begin{aligned}EV &= x(0.50) + (1-x)(-0.60)\\ &= 0.5x - 0.6 + 0.6x\\ &= 1.1x - 0.6\end{aligned}$$

The stock is better if

$$\begin{aligned}&1.1x - 0.6 > 0.045\\ \Rightarrow\ &1.1x > 0.645\\ \Rightarrow\ &x > \frac{0.645}{1.1} \approx 0.59\end{aligned}$$

13.

Outcome	*Probability*	*Value*
8 winning spots	$\frac{{}_{20}C_8}{{}_{80}C_8} \approx 0.0000043$	\$18,999
7 winning spots	$\frac{{}_{20}C_7\cdot{}_{60}C_1}{{}_{80}C_8} \approx 0.000160455$	\$1479
6 winning spots	$\frac{{}_{20}C_6\cdot{}_{60}C_2}{{}_{80}C_8} \approx 0.002366714$	\$99
5 winning spots	$\frac{{}_{20}C_5\cdot{}_{60}C_3}{{}_{80}C_8} \approx 0.0183025856$	\$4
4 winning spots	$\frac{{}_{20}C_4\cdot{}_{60}C_4}{{}_{80}C_8} \approx 0.0815037015$	\$0
$<$ 4 winning spots	$1 -$ (sum of above) ≈ 0.897662	$-\$1$

$$\begin{aligned}EV = {}&0.000004(18,999) + 0.00016(1479) + 0.002367(99) +\\ &0.018303(4) + 0.081504(0) + 0.897662(-1)\\ \approx {}&-\$0.28\end{aligned}$$

17.

Outcome	Probability	Value
Burns	0.01	120,000
Does not burn	0.99	0

$$\begin{aligned} EV &= 0.01(120,000) + 0.99(0) \\ &= 1200 \end{aligned}$$

The premium should be at least \$1200 per year.

Section 6.6
Conditional Probability

1a. $p(N) = \frac{140}{600} = \frac{7}{30}$ (probability of a "no" answer among men and women surveyed)

1b. $p(W) = \frac{320}{600} = \frac{8}{15}$ (probability that the person surveyed is a woman)

1c. $p(N|W) = \frac{p(N \cap W)}{p(W)} = \frac{\frac{45}{600}}{\frac{320}{600}} = \frac{45}{600} \cdot \frac{600}{320} = \frac{45}{320} = \frac{9}{64}$ (probability of a "no" answer given that the person questioned is a woman)

1d. $p(W|N) = \frac{p(W \cap N)}{p(N)} = \frac{\frac{45}{600}}{\frac{140}{600}} = \frac{45}{600} \cdot \frac{600}{140} = \frac{45}{140} = \frac{9}{28}$ (probability that the person questioned is a woman given that the answer is a "no")

1e. $p(N \cap W) = \frac{45}{600} = \frac{3}{40}$ (probability that the answer is "no" and the person questioned is a woman)

1f. $p(W \cap N) = \frac{45}{600} = \frac{3}{40}$ (since $N \cap W = W \cap N$, this is the same as Exercise 1e)

5. $\frac{\frac{5{,}800{,}000}{162{,}000{,}000}}{\frac{16{,}900{,}000}{162{,}000{,}000}} = \frac{5{,}800{,}000}{16{,}900{,}000} = \frac{58}{169} \approx 0.34$

9a. $p(\text{black}) = \frac{3}{12} = \frac{1}{4}$

9b. $p(\text{black} \cap \text{odd}) = \frac{2}{12} = \frac{1}{6}$

9c. $p(\text{black}|\text{odd}) = \frac{p(\text{black} \cap \text{odd})}{p(\text{odd})} = \frac{\frac{1}{6}}{\frac{1}{4}} = \frac{1}{6} \cdot \frac{4}{1} = \frac{2}{3}$

BLACK		RED	
♠ SPADES	♣ CLUBS	♥ HEARTS	♦ DIAMONDS
Ace	Ace	Ace	Ace
King	King	King	King
Queen	Queen	Queen	Queen
Jack	Jack	Jack	Jack
10	10	10	10
9	9	9	9
8	8	8	8
7	7	7	7
6	6	6	6
5	5	5	5
4	4	4	4
3	3	3	3
2	2	2	2

13a. $\frac{13}{52} = \frac{1}{4}$

13b. $\frac{13}{51}$

13c. $\frac{1}{4} \cdot \frac{13}{51} = \frac{13}{204}$

13d.

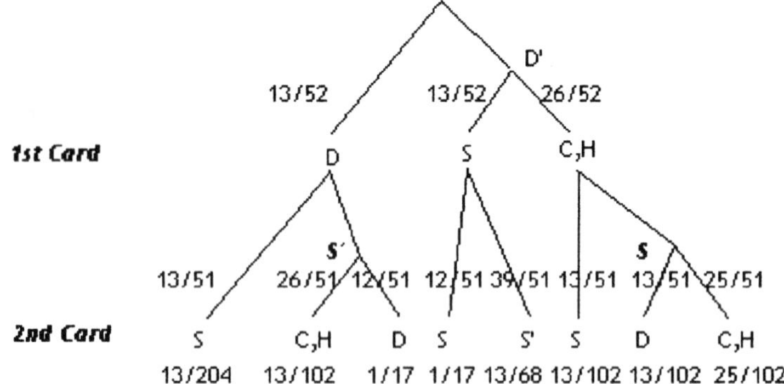

17a. $p(6) = \frac{1}{6}$

17b. $p(6|\text{even}) = \frac{p(6 \cap \text{even})}{p(\text{even})} = \frac{\frac{1}{6}}{\frac{1}{2}} = \frac{1}{3}$

17c. 0 (since the number rolled was odd, it can't be a 6)

17d. 1 (since a 6 was rolled, it is even)

21a. $\frac{3}{36} = \frac{1}{12}$

21b. Since there are 10 sums which are less than 6, and of those, 3 add to 4, the probability is $\frac{3}{10}$.

21c. 1

25.

$$\frac{n(P \cap H)}{n(H)} = \frac{125}{125 + 148}$$
$$\approx 0.46$$

Alternatively, this problem can be solved with a tree diagram. Note that of those who made a purchase, people were either happy (H) or unhappy (H'); their sum, $125 + 111 = 236$, must be the number of people who made a purchase. With this information one can construct the following tree diagram, where P indicates the purchasers and P' the non-purchasers. Note that each branch must add to 1.

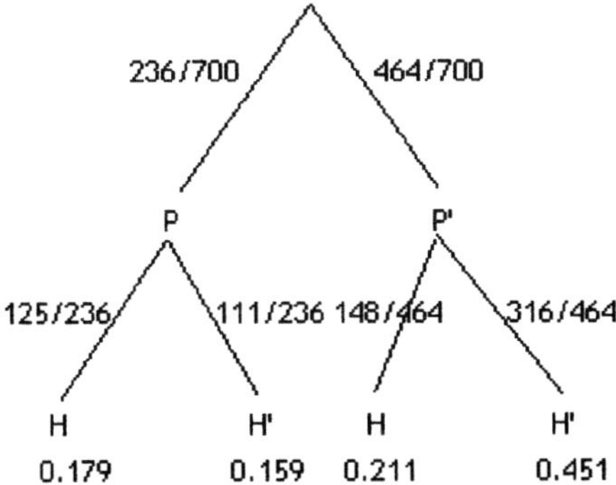

Now, the probability that a shopper who was happy with the service made a purchase is

$$\begin{aligned} p(P|H) &= \frac{p(P\cap H)}{p(H)} \\ &= \frac{0.179}{0.179+0.211} \\ &= 0.459 \end{aligned}$$

Perhaps the merchandise is not as satisfactory as it could be, or the prices may be too high.

29. We use a tree diagram:

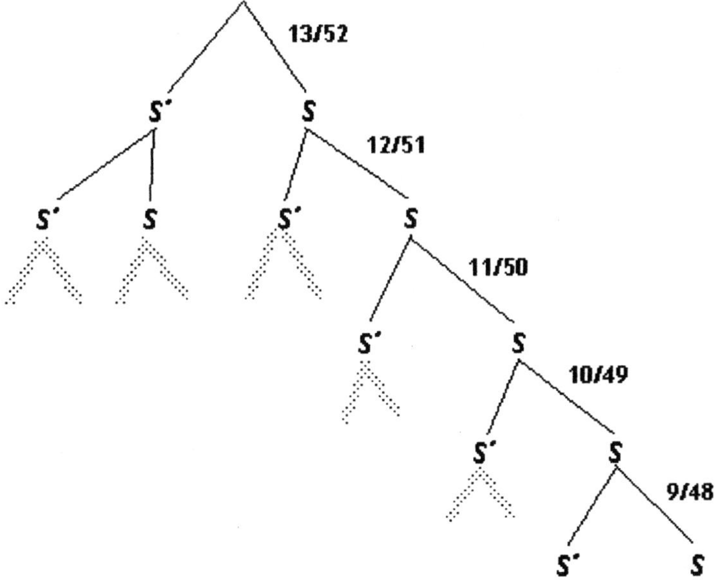

(Only the necessary part of the tree is drawn.) Start at the condition "2nd is a spade" and multiply:

$$\frac{12}{51} \cdot \frac{11}{50} \cdot \frac{10}{49} \cdot \frac{9}{48} \approx 0.0020.$$

33.

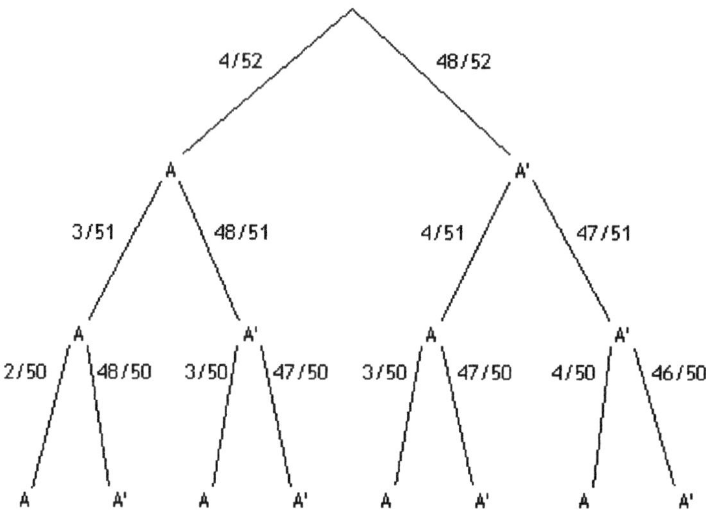

1/5525 24/5525 24/5525 376/5525 24/5525 376/5525 376/5525 4324/5525

$\frac{376}{5525} + \frac{376}{5525} + \frac{376}{5525} = \frac{1128}{5525}$

37-39, tree diagram:

37.

$$\begin{aligned} p(OK' \cap J) &= p(OK'|J) \cdot p(J) \\ &= 0.017 \cdot 0.38 \\ &= 0.00646 \end{aligned}$$

41-43, tree diagram:

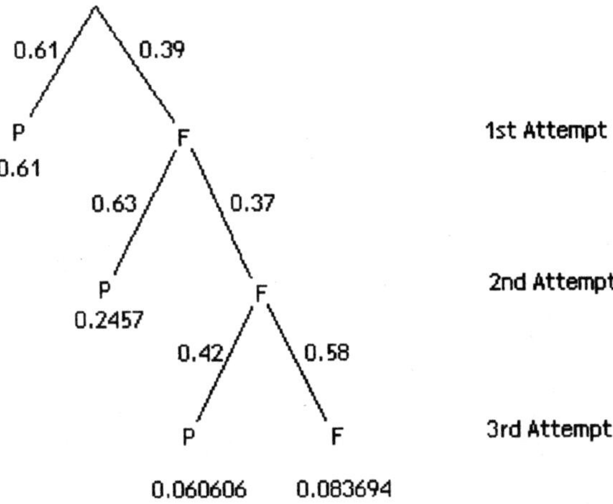

41.

$$\begin{aligned} p(\text{pass exam}) &= 1 - p(\text{fail 3 times}) \\ &= 1 - 0.083694 \\ &\approx 0.92 \end{aligned}$$

45.

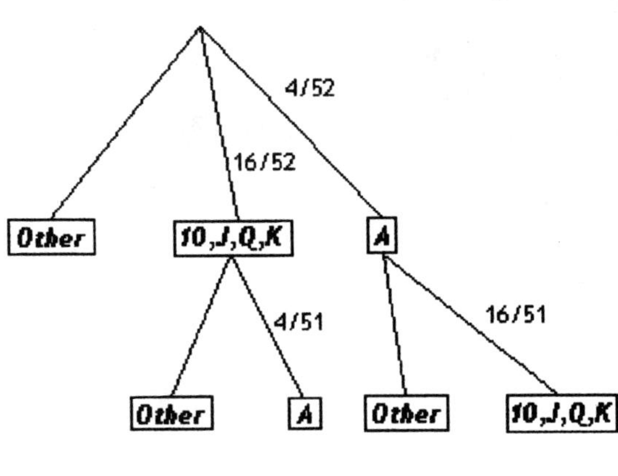

$$\frac{4}{52} \cdot \frac{16}{51} + \frac{16}{52} \cdot \frac{4}{51} \approx 0.05$$

49.

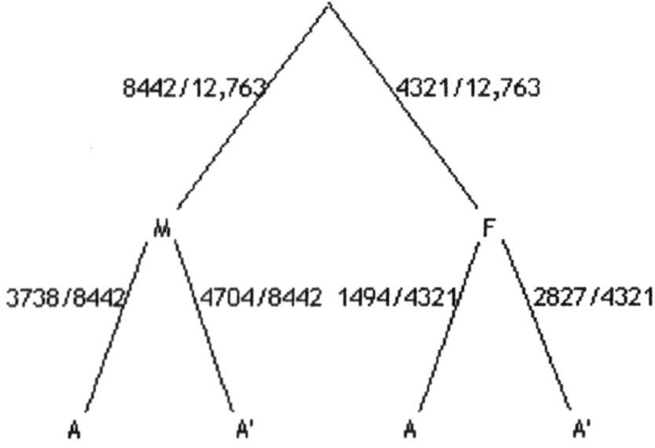

49a. $\frac{3738+1494}{12,763} = \frac{5232}{12,763}$

49b. $\frac{3738}{8442}$

49c. $\frac{1494}{4321}$

49d. Yes.

53a. $p(\text{yes or don't know}|W) = \frac{256+19}{320} = \frac{55}{64}$

53b. $p(N|M) = \frac{95}{280} = \frac{19}{56}$

53c. $p(\text{yes or don't know}|M) = \frac{162+23}{280} = \frac{37}{56}$

53d. $N'|W$

Section 6.7 Independence

1a. Nowadays these are approaching independent status; in the past there was much discrimination against women becoming doctors, so these were dependent.

1b. These are not mutually exclusive.

5a. Independent.

5b. Not mutually exclusive, given aging.

9a. $p(E) = \frac{1}{6}$; $p(E|F) = 0$. Dependent.

9b. Mutually exclusive.

13. $p(P|H) \approx 0.46$. Also $p(P) = \frac{125+111}{125+111+148+316} \approx 0.34$. Dependent. People who are happy with the service are more likely to make a purchase.

17a. Since the events are independent, the probability of failure of all three systems is $(0.01)^3 = 0.000001 = \frac{1}{1{,}000{,}000}$.

17b. At least 4; 3 gives a probability of total failure $\frac{1}{100{,}000{,}000} > \frac{1}{1{,}000{,}000{,}000}$.

1a. The conditions of the schoolchildren *healthy* (H) and *unhealthy* (H'), i.e., afflicted with white lung disease, are mutually exclusive events whose union forms the sample space, so we may use Bayes' Theorem to find the probability that a schoolchild is unhealthy given a positive test result, or $p(H'|+)$:

$$\begin{aligned} p(H'|+) &= \frac{p(+|H') \cdot p(H')}{p(+|H') \cdot p(H') + p(+|H) \cdot p(H)} \\ &= \frac{(0.99)(0.001)}{(0.99)(0.001) + (0.02)(0.999)} \\ &= \frac{0.00099}{0.02097} \\ &\approx 0.0472 \end{aligned}$$

1b.

$$\begin{aligned} p(H|-) &= \frac{p(-|H) \cdot p(H)}{p(-|H) \cdot p(H) + p(-|H') \cdot p(H')} \\ &= \frac{(0.98)(0.999)}{(0.98)(0.999) + (0.01)(0.001)} \\ &= \frac{0.97902}{0.97903} \\ &\approx 0.99999 \end{aligned}$$

1c.

$$\begin{aligned} p(H|+) &= \frac{p(+|H) \cdot p(H)}{p(+|H) \cdot p(H) + p(+|H') \cdot p(H')} \\ &= \frac{(0.02)(0.999)}{(0.02)(0.999) + (0.99)(0.001)} \\ &= \frac{0.01998}{0.02097} \\ &\approx 0.9528 \end{aligned}$$

1d.

$$\begin{aligned} p(H'|-) &= \frac{p(-|H') \cdot p(H')}{p(-|H') \cdot p(H') + p(-|H) \cdot p(H)} \\ &= \frac{(0.01)(0.001)}{(0.01)(0.001) + (0.98)(0.999)} \\ &= \frac{0.00001}{0.97903} \\ &\approx 0.00001 \end{aligned}$$

1e. A false positive occurs when the test result is positive (i.e., indicating sickness is present) but the person is actually healthy; a false negative occurs when the test result is negative (i.e., indicating no sickness is present) but the person is actually sick.

1f. (a) and (c).

5a.

$$\begin{aligned} p(K|OK') &= \frac{p(OK'|K)\cdot p(K)}{p(OK'|K)\cdot p(K) + p(OK'|P)\cdot p(P)} \\ &= \frac{(0.04)(0.6)}{(0.04)(0.6) + (0.05)(0.4)} \\ &= \frac{0.024}{0.044} \\ &\approx 0.5455 \end{aligned}$$

5b.

$$\begin{aligned} p(P|OK') &= \frac{p(OK'|P)\cdot p(P)}{p(OK'|P)\cdot p(P) + p(OK'|K)\cdot p(K)} \\ &= \frac{(0.05)(0.4)}{(0.05)(0.4) + (0.04)(0.6)} \\ &= \frac{0.02}{0.044} \\ &\approx 0.4545 \end{aligned}$$

5c.

$$\begin{aligned} p(P|OK) &= \frac{p(OK|P)\cdot p(P)}{p(OK|P)\cdot p(P) + p(OK|K)\cdot p(K)} \\ &= \frac{(0.95)(0.4)}{(0.95)(0.4) + (0.96)(0.6)} \\ &= \frac{0.38}{0.956} \\ &\approx 0.3975 \end{aligned}$$

9.

$$\begin{aligned} p(A_k|X) &= \frac{p(X|A_k)\cdot p(A_k)}{\sum_{j=1}^{3} p(X|A_j)\cdot p(A_j)} \\ &= \frac{p(X|A_k)\cdot p(A_k)}{p(X|A_1)\cdot p(A_1) + p(X|A_2)\cdot p(A_2) + p(X|A_3)\cdot p(A_3)}, \end{aligned}$$

where X represents C or D.

13a. ("Y" means a "yes" vote; "N" means a "no" vote.)

$$\begin{aligned} p(R|Y) &= \frac{p(Y|R)\cdot p(R)}{p(Y|R)\cdot p(R) + p(Y|D)\cdot p(D) + p(Y|I)\cdot p(I)} \\ &= \frac{(0.31)(0.38)}{(0.31)(0.38) + (0.76)(0.51) + (0.79)(0.11)} \\ &= \frac{0.1178}{0.5923} \\ &\approx 0.1989 \end{aligned}$$

13b.

$$\begin{aligned}
p(D|Y) &= \frac{p(Y|D)\cdot p(D)}{p(Y|R)\cdot p(R) + p(Y|D)\cdot p(D) + p(Y|I)\cdot p(I)} \\
&= \frac{(0.76)(0.51)}{(0.31)(0.38) + (0.76)(0.51) + (0.79)(0.11)} \\
&= \frac{0.3876}{0.5923} \\
&\approx 0.6544
\end{aligned}$$

13c.

$$\begin{aligned}
p(R|N) &= \frac{p(N|R)\cdot p(R)}{p(N|R)\cdot p(R) + p(N|D)\cdot p(D) + p(N|I)\cdot p(I)} \\
&= \frac{(0.69)(0.38)}{(0.69)(0.38) + (0.24)(0.51) + (0.21)(0.11)} \\
&= \frac{0.2622}{0.4077} \\
&\approx 0.6431
\end{aligned}$$

Chapter 6
Review Exercises

1a. $\frac{1}{2}$; 1:1

1b. $\frac{4}{52} = \frac{1}{13}$; 1:12

1c. $\frac{13}{52} = \frac{1}{4}$; 1:3

1d. $\frac{1}{52}$; 1:51

1e.

$$\begin{aligned} p(\text{queen} \cup \text{club}) &= p(\text{queen}) + p(\text{club}) - p(\text{queen} \cap \text{club}) \\ &= \frac{1}{13} + \frac{1}{4} - \frac{1}{52} \\ &= \frac{4}{13} \end{aligned}$$

Odds are 4:9.

1f. $1 - \frac{1}{13} = \frac{12}{13}$; odds are 12:1.

2a. The actual coin tossing.

2b.
$S = \{(H,H,H),(H,H,T),(H,T,H),(T,H,H),(H,T,T),(T,H,T),(T,T,H),(T,T,T)\}$

2c. $E = \{(H,T,T),(T,H,T),(T,T,H)\}$

2d. $F = \{(H,T,T),(T,H,T),(T,T,H),(T,T,T)\}$

2e. $p(E) = \frac{3}{8}$; $o(E) = 3:5$

2f. $p(F) = \frac{4}{8} = \frac{1}{2}$; $o(F) = 1:1$

3a. $\frac{6}{36} = \frac{1}{6}$

3b. $\frac{2}{36} = \frac{1}{18}$

3c. $\frac{1}{6} + \frac{1}{18} + \frac{1}{6} = \frac{7}{18}$

3d. $\frac{6}{36} = \frac{1}{6}$

3e. $\frac{22}{36} = \frac{11}{18}$

3f. $1 - \frac{11}{18} = \frac{7}{18}$

4a. $\frac{13}{52} \cdot \frac{12}{51} \cdot \frac{11}{50} = \frac{11}{850}$

4b. $3 \cdot \frac{39 \cdot 13 \cdot 12}{52 \cdot 51 \cdot 50} = \frac{117}{850}$

4c. $\frac{117}{850} + \frac{11}{850} = \frac{64}{425}$

4d. $\frac{1}{52} \cdot \frac{1}{51} \cdot \frac{1}{50} = \frac{1}{132{,}600}$

5a. $\left(\frac{1}{6}\right)^3 = \frac{1}{216}$

5b. $3\left(\frac{1}{6} \cdot \frac{1}{6} \cdot \frac{5}{6}\right) = \frac{5}{72}$

5c. $\frac{1}{216} + \frac{5}{72} = \frac{2}{27}$

6a. $0.08 + 0.06 = 0.14$

6b. No, since the probabilities that "a plant must be returned" and "a plant must be returned given it was from Herb's Herbs" are different.

7a.

	S	s
s	Ss	ss
s	Ss	ss

p(long-stemmed) $= \frac{1}{2}$

7b. p(short-stemmed) $= \frac{1}{2}$

8a.

	C	c
C	CC	Cc
c	Cc	cc

p(child will have cystic fibrosis) $= \frac{1}{4}$

8b. p(child will be a carrier) $= \frac{1}{2}$

8c. p(child will neither have the disease nor be a carrier) $= \frac{1}{4}$

9a.

	S	s
S	SS	Ss
s	Ss	ss

$\frac{1}{4}$

9b. $\frac{1}{2}$

9c. $\frac{1}{4}$

10a.

	H	h
h	Hh	hh
h	Hh	hh

$\frac{1}{2}$

10b. $\frac{1}{2}$

11a.

	T	t
T	TT	Tt
T	TT	Tt

0

11b. $\frac{1}{2}$

11c. $\frac{1}{2}$

12a. $\frac{266}{472}, \frac{184}{472}$

12b. $\frac{131}{328}, \frac{181}{328}$

12c. $\frac{266}{397}, \frac{131}{397}$

12d. $\frac{184}{365}, \frac{181}{365}$

12e. urban; urban

12f. O'Neill; Bell

12g. No

12h. O'Neill

13a. Dependent; not mutually exclusive

13b. Independent; not mutually exclusive.

13c. Dependent; mutually exclusive.

13d. Dependent; not mutually exclusive.

13e. Independent; not mutually exclusive.

14a.

$$\begin{aligned} p(AR|OK') &= \frac{p(OK'|AR)\,p(AR)}{p(OK'|AR)\,p(AR) + p(OK'|NV)\,p(NV)} \\ &= \frac{(0.02)(0.39)}{(0.02)(0.39) + (0.017)(0.61)} \\ &= \frac{0.0078}{0.01817} \\ &\approx 0.4293 \end{aligned}$$

14b.

$$\begin{aligned} p(NV|OK') &= \frac{p(OK'|NV)p(NV)}{p(OK'|AR)p(AR)+p(OK'|NV)p(NV)} \\ &= \frac{(0.017)(0.61)}{(0.02)(0.39)+(0.017)(0.61)} \\ &= \frac{0.01037}{0.01817} \\ &\approx 0.5707 \end{aligned}$$

15a. In 9-spot Keno, for example, there are ${}_{80}C_9$ ways of marking a 9-spot card. With 9 specific numbers marked and 20 numbers randomly drawn, there are ${}_{20}C_9$ (out of the ${}_{80}C_9$) ways for all of the 9 specifically marked numbers to be present among the 20 numbers drawn. Hence the probability that all 9 marked numbers are drawn is

$$\frac{{}_{20}C_9}{{}_{80}C_9} \approx 0.000000724277.$$

Outcome	*Probability*
9 winning spots	$\frac{{}_{20}C_9}{{}_{80}C_9} \approx 0.000000724277$
8 winning spots	$\frac{{}_{20}C_8 \cdot {}_{60}C_1}{{}_{80}C_9} \approx 0.00003259245$
7 winning spots	$\frac{{}_{20}C_7 \cdot {}_{60}C_2}{{}_{80}C_9} \approx 0.0005916784$
6 winning spots	$\frac{{}_{20}C_6 \cdot {}_{60}C_3}{{}_{80}C_9} \approx 0.005719558$
5 winning spots	$\frac{{}_{20}C_5 \cdot {}_{60}C_4}{{}_{80}C_9} \approx 0.0326014806$
4 winning spots	$\frac{{}_{20}C_4 \cdot {}_{60}C_5}{{}_{80}C_9} \approx 0.114105182$
$<$ 4 winning spots	$1 -$ (sum of above) ≈ 0.846949

15b. Expected value $= (0.00000072)(25,000) + (0.0000326)(6000) + (0.0005917)(390) + (0.00572)(50) + (0.0326)(1) + (0.1141 + 0.847)(-1) = -\$0.20$

16. Some bets in craps are better.

17. The probability that the event B occurs given that the event A occurs is the conditional probability of B given A.

18. If $p(B|A) = p(B)$ we say that A and B are independent.

19.

$$\begin{aligned} &0 \le n(E) \le n(5) \\ &\Rightarrow \frac{0}{n(5)} < \frac{n(E)}{n(5)} < \frac{n(5)}{n(5)} \\ &\Rightarrow 0 < p(E) < 1 \end{aligned}$$

22. Pascal, Fermat, and Cardano in answer to a question involving bets in a game of chance.

Chapter 7 Markov Chains

Section 7.1 Introduction to Markov Chains

1a. p(current purchase is KickKola) $= 0.14$; p(current purchase is not KickKola) $= 1 - 0.14 = 0.86$

1b. $P = [0.14 \quad 0.86]$

5a. p(Silver's Gym) $= 0.48$; p(Fitness Lab) $= 0.37$; p(ThinNFit) $= 1 - (0.48 + 0.37) = 0.15$

5b. $P = [0.48 \quad 0.37 \quad 0.15]$

9a. Let R represent that they plan to rent.

$$\begin{aligned} p(R' \text{ next}|R \text{ now}) &= 0.12 \\ p(R \text{ next}|R \text{ now}) &= 1 - 0.12 = 0.88 \\ p(R \text{ next}|R' \text{ now}) &= 0.03 \\ p(R' \text{ next}|R' \text{ now}) &= 1 - 0.03 = 0.97 \end{aligned}$$

9b.

$$\begin{aligned} T &= \begin{bmatrix} p(R|R) & p(R'|R) \\ p(R|R') & p(R'|R') \end{bmatrix} \\ &= \begin{bmatrix} 0.88 & 0.12 \\ 0.03 & 0.97 \end{bmatrix} \end{aligned}$$

13a.

$$\begin{aligned} PT &= [0.14 \quad 0.86] \begin{bmatrix} 0.63 & 0.37 \\ 0.12 & 0.88 \end{bmatrix} \\ &= [0.1914 \quad 0.8086] \end{aligned}$$

After the first following purchase, KickKola's market share will be 19.14%.

13b.

$$\begin{aligned} PT^2 = (PT)T &= [0.1914 \quad 0.8086] \begin{bmatrix} 0.63 & 0.37 \\ 0.12 & 0.88 \end{bmatrix} \\ &= [0.217614 \quad 0.782386] \end{aligned}$$

After the second following purchase, KickKola's market share will be about 22%.

17.

$$P = [0.41 \quad 0.59]$$

$p(S.C.\|S.C.) = 0.12$	$p(S.C.'\|S.C.) = 1 - 0.12 = 0.88$
$p(S.C.\|S.C.') = 0.31$	$p(S.C.'\|S.C.') = 1 - 0.31 = 0.69$

$$T = \begin{bmatrix} 0.12 & 0.88 \\ 0.31 & 0.69 \end{bmatrix}$$

$$\begin{aligned} PT^2 &= \begin{bmatrix} 0.41 & 0.59 \end{bmatrix} \begin{bmatrix} 0.12 & 0.88 \\ 0.31 & 0.69 \end{bmatrix}^2 \\ &= \begin{bmatrix} 0.41 & 0.59 \end{bmatrix} \begin{bmatrix} 0.2872 & 0.7128 \\ 0.2511 & 0.7489 \end{bmatrix} \\ &= \begin{bmatrix} 0.265901 & 0.734099 \end{bmatrix} \end{aligned}$$

In four years Sierra Cruiser's market share will be about 26.59%.

21a. (Numbers are millions of acres.) Column 1 is cropland (C), column 2 is potential cropland (PC), and column 3 is noncropland (NC):

$$\begin{aligned} P &= \begin{bmatrix} \frac{413}{413+127+856} = \frac{413}{1396} & \frac{127}{1396} & \frac{856}{1396} \end{bmatrix} \\ &\approx \begin{bmatrix} 0.296 & 0.091 & 0.613 \end{bmatrix} \end{aligned}$$

21b.

$p(C\|C) = \frac{413-34}{431} \approx 0.879$	$p(PC\|C) = \frac{17}{431} \approx 0.039$	$p(NC\|C) = \frac{35}{431} \approx 0.081$
$p(C\|PC) = \frac{34}{144} \approx 0.236$	$p(PC\|PC) = \frac{127-17}{144} \approx 0.764$	$p(NC\|PC) = 0$
$p(C\|NC) = 0$	$p(PC\|NC) = 0$	$p(NC\|NC) = \frac{821}{821}$

$$\therefore T = \begin{bmatrix} 0.879 & 0.039 & 0.081 \\ 0.236 & 0.764 & 0 \\ 0 & 0 & 1 \end{bmatrix}$$

21c. The 1977 totals are (millions of acres): cropland, 413; potential cropland, 127; non-cropland, 856. We put this data into a matrix and multiply by the transition matrix T:

$$\begin{aligned} \begin{bmatrix} 413 & 127 & 856 \end{bmatrix} \begin{bmatrix} 0.879 & 0.039 & 0.081 \\ 0.236 & 0.764 & 0 \\ 0 & 0 & 1 \end{bmatrix} &= \begin{bmatrix} 392.999 & 113.135 & 889.453 \end{bmatrix} \\ &\approx \begin{bmatrix} 393 & 113 & 889 \end{bmatrix} \end{aligned}$$

Thus in 1987 there should be about 393 million acres of cropland, 113 million acres of potential cropland, and 889 million acres of non-cropland.

21d. We use the 1987 data above in a similar way.

$$\begin{aligned} \begin{bmatrix} 393 & 113 & 889 \end{bmatrix} \begin{bmatrix} 0.879 & 0.039 & 0.081 \\ 0.236 & 0.764 & 0 \\ 0 & 0 & 1 \end{bmatrix} &= \begin{bmatrix} 372.115 & 101.659 & 920.883 \end{bmatrix} \\ &\approx \begin{bmatrix} 372 & 102 & 921 \end{bmatrix} \end{aligned}$$

Thus in 1997 there should be about 372 million acres of cropland, 102 million acres of potential cropland, and 921 million acres of non-cropland.

Regular Markov Chains

1.

$$LT = L$$

$$[x \quad y]\begin{bmatrix} 0.1 & 0.9 \\ 0.2 & 0.8 \end{bmatrix} = [x \quad y]$$

$$[0.1x + 0.2y \quad 0.9x + 0.8y] = [x \quad y]$$

$$0.1x + 0.2y = x,$$
$$0.9x + 0.8y = y$$

$$-0.9x + 0.2y = 0,$$
$$0.9x - 0.2y = 0,$$
$$x + y = 1$$

Using Gauss-Jordan or a calculator on the matrix

$$\begin{bmatrix} -0.9 & 0.2 & 0 \\ 0.9 & -0.2 & 0 \\ 1 & 1 & 1 \end{bmatrix},$$

we obtain

$$\begin{bmatrix} 1 & 0 & 0.181818 \\ 0 & 1 & 0.818182 \\ 0 & 0 & 0 \end{bmatrix}.$$

Thus

$$L = [x \quad y] = [0.181818 \quad 0.818182].$$

5. From Exercise 13, Sec. 7.1, we have $T = \begin{bmatrix} 0.63 & 0.37 \\ 0.12 & 0.88 \end{bmatrix}$.

$$LT = L$$

$$[x \quad y]\begin{bmatrix} 0.63 & 0.37 \\ 0.12 & 0.88 \end{bmatrix} = [x \quad y]$$

$$[0.63x + 0.12y \quad 0.37x + 0.88y] = [x \quad y]$$

$$0.63x + 0.12y = x,$$
$$0.37x + 0.88y = y$$

$$-0.37x + 0.12y = 0,$$
$$0.37x - 0.12y = 0,$$
$$x + y = 1$$

Using Gauss-Jordan or a calculator on the matrix

$$\begin{bmatrix} -0.37 & 0.12 & 0 \\ 0.37 & -0.12 & 0 \\ 1 & 1 & 1 \end{bmatrix},$$

we obtain

$$\begin{bmatrix} 1 & 0 & 0.244897959184 \\ 0 & 1 & 0.755102040816 \\ 0 & 0 & 0 \end{bmatrix}.$$

Thus

$$L = [\,x \quad y\,] \approx [\,0.2449 \quad 0.7551\,].$$

In other words, if current trends continue, KickKola's market share will stabilize at about 24.49%.

9. From Exercise 9b we have that $T = \begin{bmatrix} 0.88 & 0.12 \\ 0.03 & 0.97 \end{bmatrix}$.

$$LT = L$$

$$[\,x \quad y\,]\begin{bmatrix} 0.88 & 0.12 \\ 0.03 & 0.97 \end{bmatrix} = [\,x \quad y\,]$$

$$[\,0.88x + 0.03y \quad 0.12x + 0.97y\,] = [\,x \quad y\,]$$

$$\begin{aligned} 0.88x + 0.03y &= x, \\ 0.12x + 0.97y &= y \end{aligned}$$

$$\begin{aligned} -0.12x + 0.03y &= 0, \\ 0.12x - 0.03y &= 0, \\ x + y &= 1 \end{aligned}$$

Using Gauss-Jordan or a calculator on the matrix

$$\begin{bmatrix} -0.12 & 0.03 & 0 \\ 0.12 & -0.03 & 0 \\ 1 & 1 & 1 \end{bmatrix},$$

we obtain

$$\begin{bmatrix} 1 & 0 & 0.2 \\ 0 & 1 & 0.8 \\ 0 & 0 & 0 \end{bmatrix}.$$

Thus

$$L = [\,x \quad y\,] = [\,0.2 \quad 0.8\,].$$

In other words, if current trends continue, 20% of the population of Metropolis will rent and 80% will own their own homes.

13.

$$LT = L$$

$$[\,x_1 \quad x_2 \quad x_3 \quad x_4 \quad x_5 \quad x_6\,]\begin{bmatrix} 0.6 & 0.1 & 0.12 & 0.08 & 0.07 & 0.03 \\ 0.09 & 0.64 & 0.07 & 0.08 & 0.09 & 0.03 \\ 0.19 & 0.1 & 0.54 & 0.09 & 0.07 & 0.01 \\ 0.05 & 0.06 & 0.05 & 0.69 & 0.12 & 0.03 \\ 0.03 & 0.05 & 0.13 & 0.16 & 0.61 & 0.02 \\ 0.18 & 0.15 & 0.18 & 0.16 & 0.17 & 0.16 \end{bmatrix} = [\,x_1 \quad x_2 \quad x_3 \quad x_4 \quad x_5 \quad x_6\,]$$

$$[\,0.6x_1 + 0.09x_2 + 0.19x_3 + 0.05x_4 + 0.03x_5 + 0.18x_6 \quad \cdots\,] = [\,x_1 \quad x_2 \quad x_3 \quad x_4 \quad x_5 \quad x_6\,]$$

$$
\begin{aligned}
0.6x_1 + 0.09x_2 + 0.19x_3 + 0.05x_4 + 0.03x_5 + 0.18x_6 &= x_1, \\
0.1x_1 + 0.64x_2 + 0.1x_3 + 0.06x_4 + 0.05x_5 + 0.15x_6 &= x_2, \\
0.12x_1 + 0.07x_2 + 0.54x_3 + 0.05x_4 + 0.13x_5 + 0.18x_6 &= x_3, \\
0.08x_1 + 0.08x_2 + 0.09x_3 + 0.69x_4 + 0.16x_5 + 0.16x_6 &= x_4, \\
0.07x_1 + 0.09x_2 + 0.07x_3 + 0.12x_4 + 0.61x_5 + 0.17x_6 &= x_5, \\
0.03x_1 + 0.03x_2 + 0.01x_3 + 0.03x_4 + 0.02x_5 + 0.16x_6 &= x_6
\end{aligned}
$$

$$
\begin{aligned}
-0.4x_1 + 0.09x_2 + 0.19x_3 + 0.05x_4 + 0.03x_5 + 0.18x_6 &= 0, \\
0.1x_1 - 0.36x_2 + 0.1x_3 + 0.06x_4 + 0.05x_5 + 0.15x_6 &= 0, \\
0.12x_1 + 0.07x_2 - 0.46x_3 + 0.05x_4 + 0.13x_5 + 0.18x_6 &= 0, \\
0.08x_1 + 0.08x_2 + 0.09x_3 - 0.31x_4 + 0.16x_5 + 0.16x_6 &= 0, \\
0.07x_1 + 0.09x_2 + 0.07x_3 + 0.12x_4 - 0.39x_5 + 0.17x_6 &= 0, \\
0.03x_1 + 0.03x_2 + 0.01x_3 + 0.03x_4 + 0.02x_5 - 0.84x_6 &= 0
\end{aligned}
$$

Using Gauss-Jordan or a calculator on the matrix

$$
\begin{bmatrix}
-0.4 & 0.09 & 0.19 & 0.05 & 0.03 & 0.18 & 0 \\
0.1 & -0.36 & 0.1 & 0.06 & 0.05 & 0.15 & 0 \\
0.12 & 0.07 & -0.46 & 0.05 & 0.13 & 0.18 & 0 \\
0.08 & 0.08 & 0.09 & -0.31 & 0.16 & 0.16 & 0 \\
0.07 & 0.09 & 0.07 & 0.12 & -0.39 & 0.17 & 0 \\
0.03 & 0.03 & 0.01 & 0.03 & 0.02 & -0.84 & 0 \\
1 & 1 & 1 & 1 & 1 & 1 & 1
\end{bmatrix},
$$

we obtain

$$
\begin{bmatrix}
1 & 0 & 0 & 0 & 0 & 0 & 0.178456031458 \\
0 & 1 & 0 & 0 & 0 & 0 & 0.177244351462 \\
0 & 0 & 1 & 0 & 0 & 0 & 0.167122442383 \\
0 & 0 & 0 & 1 & 0 & 0 & 0.254990209361 \\
0 & 0 & 0 & 0 & 1 & 0 & 0.193773382198 \\
0 & 0 & 0 & 0 & 0 & 1 & 0.0284135831384 \\
0 & 0 & 0 & 0 & 0 & 0 & 0
\end{bmatrix}.
$$

Thus

$$
L \approx [\,0.1785 \quad 0.1772 \quad 0.1671 \quad 0.2550 \quad 0.1938 \quad 0.0284\,].
$$

Absorbing Markov Chains

1a. The six states are *freshman, sophomore, junior, senior, graduated,* and *quit* (in that order in the matrix).

$$T = \begin{bmatrix} 0.08 & 0.85 & 0 & 0 & 0 & 0.07 \\ 0 & 0.07 & 0.87 & 0 & 0 & 0.06 \\ 0 & 0 & 0.06 & 0.89 & 0 & 0.05 \\ 0 & 0 & 0 & 0.05 & 0.91 & 0.04 \\ 0 & 0 & 0 & 0 & 1 & 0 \\ 0 & 0 & 0 & 0 & 0 & 1 \end{bmatrix}$$

1b. It is not a regular matrix because it is not possible to go from some states to other states, such as from *graduated* to *freshman* states. This is an absorbing matrix because it has at least one absorbing state (*graduated* or *quit*), and it is possible to go from any non-absorbing state (*freshman, sophomore, junior, senior*) to an absorbing state (*graduated, quit*).

1c.

$$T = \begin{bmatrix} 0.08 & 0.85 & 0 & 0 & 0 & 0.07 \\ 0 & 0.07 & 0.87 & 0 & 0 & 0.06 \\ 0 & 0 & 0.06 & 0.89 & 0 & 0.05 \\ 0 & 0 & 0 & 0.05 & 0.91 & 0.04 \\ 0 & 0 & 0 & 0 & 1 & 0 \\ 0 & 0 & 0 & 0 & 0 & 1 \end{bmatrix},$$

$$T^2 = \begin{bmatrix} 0.0064 & 0.1275 & 0.7395 & 0 & 0 & 0.1266 \\ 0 & 0.0049 & 0.1131 & 0.7743 & 0 & 0.1077 \\ 0 & 0 & 0.0036 & 0.0979 & 0.8099 & 0.0886 \\ 0 & 0 & 0 & 0.0025 & 0.9555 & 0.042 \\ 0 & 0 & 0 & 0 & 1 & 0 \\ 0 & 0 & 0 & 0 & 0 & 1 \end{bmatrix}.$$

Row 1, column 1 of T shows that 8% of entering freshmen return as freshmen. Row 1, column 1 of T^2 shows that 0.64% of the second year freshmen spend their third year as freshmen. The expected value of the number of years a freshman spends as a freshman (estimate) is

$$(1.00)(1) + (0.08)(1) + (0.0064)(1) = 1.0864.$$

1d. Recall that $(I - N)^{-1}$ is the fundamental matrix:

$$(I-N)^{-1} = \left(\begin{bmatrix} 1 & 0 & 0 & 0 \\ 0 & 1 & 0 & 0 \\ 0 & 0 & 1 & 0 \\ 0 & 0 & 0 & 1 \end{bmatrix} - \begin{bmatrix} 0.08 & 0.85 & 0 & 0 \\ 0 & 0.07 & 0.87 & 0 \\ 0 & 0 & 0.06 & 0.89 \\ 0 & 0 & 0 & 0.05 \end{bmatrix} \right)^{-1}$$

$$= \begin{bmatrix} 0.92 & -0.85 & 0 & 0 \\ 0 & 0.93 & -0.87 & 0 \\ 0 & 0 & 0.94 & -0.89 \\ 0 & 0 & 0 & 0.95 \end{bmatrix}^{-1}$$

$$\approx \begin{bmatrix} 1.08695652174 & 0.993454885456 & 0.919474202498 & 0.861402147607 \\ 0 & 1.0752688172 & 0.995195607409 & 0.932341147999 \\ 0 & 0 & 1.06382978723 & 0.996640537516 \\ 0 & 0 & 0 & 1.05263157895 \end{bmatrix}$$

From row 1, column 1 of $(I-N)^{-1}$ we see that on the average a freshman spends about 1.09 years as a freshman.

1e. From $(I-N)^{-1}$ a beginning freshman can expect to spend

$$1.08695652174 + 0.993454885456 + 0.919474202498 + 0.861402147607 \approx 3.86 \text{ years}$$

at the university.

1f. From $(I-N)^{-1}$:

$$1.06382978723 + 0.996640537516 \approx 2.06 \text{ yrs.}$$

1g.

$$T^5 = \begin{bmatrix} 0.0000032768 & 0.0001356685 & 0.0025623675 & 0.0279715875 & 0.754640523 & 0.2146865767 \\ 0 & 0.0000016807 & 0.0000785697 & 0.001695717 & 0.846334421 & 0.1514905905 \\ 0 & 0 & 0.0000007776 & 0.0000413939 & 0.9069025329 & 0.0930552956 \\ 0 & 0 & 0 & 0.0000003125 & 0.9578944375 & 0.04210525 \\ 0 & 0 & 0 & 0 & 1 & 0 \end{bmatrix}$$

Note that column 5 is the *graduated* column while row 1 is the *freshman* row. Thus about 75% of the freshmen have graduated after 5 years, and given enough time, an additional 2.8% + 0.3% = 3.1% could still graduate; thus between 75% and 78.1% of the freshmen will graduate.

1h. We need the matrix $(I-N)^{-1}A$, where A is the matrix formed by deleting the rows of T corresponding to the absorbing states, viz. the last two rows, and then forming A from the absorbing columns, viz. the last two columns:

$$A = \begin{bmatrix} 0 & 0.07 \\ 0 & 0.06 \\ 0 & 0.05 \\ 0.91 & 0.04 \end{bmatrix}.$$

Thus

$$(I-N)^{-1}A = \begin{bmatrix} 1.08695652174 & 0.993454885456 & 0.919474202498 & 0.861402147607 \\ 0 & 1.0752688172 & 0.995195607409 & 0.932341147999 \\ 0 & 0 & 1.06382978723 & 0.996640537516 \\ 0 & 0 & 0 & 1.05263157895 \end{bmatrix} \begin{bmatrix} 0 & 0.07 \\ 0 & 0.06 \\ 0 & 0.05 \\ 0.91 & 0.04 \end{bmatrix}$$

$$= \begin{bmatrix} 0.783875954322 & 0.216124045678 \\ 0.848430444679 & 0.151569555322 \\ 0.90694288914 & 0.0930571108621 \\ 0.957894736845 & 0.042105263158 \end{bmatrix}$$

Rows 1 through 4 are the non-absorbing states and columns 1 and 2 are the absorbing states. From row 1, column 1 the probability that an entering freshman will graduate is about 78.39%.

1i. The probability that a junior will graduate is obtained from row 3, column 1 of $(I-N)^{-1}A$, viz. about 90.69%.

5a. The non-absorbing states are *new, 1 month overdue, 2 months overdue, 3 months overdue*, and the absorbing states are *paid, bad*. T is written in that order.

$$T = \begin{bmatrix} 0 & 0.3 & 0 & 0 & 0.7 & 0 \\ 0 & 0 & 0.4 & 0 & 0.6 & 0 \\ 0 & 0 & 0 & 0.5 & 0.5 & 0 \\ 0 & 0 & 0 & 0 & 0.4 & 0.6 \\ 0 & 0 & 0 & 0 & 1 & 0 \\ 0 & 0 & 0 & 0 & 0 & 1 \end{bmatrix}$$

5b. It is not a regular matrix because it is not possible to go from some states to other states, such as from *paid* to overdue states. This is an absorbing matrix because it has at least one absorbing state (*paid* or *bad*), and it is possible to go from any non-absorbing state (*new, 1 month overdue, 2 months overdue, 3 months overdue*) to an absorbing state (*paid, bad*).

5c. Recall that N is the matrix which contains only the non-absorbing states:

$$N = \begin{bmatrix} 0 & 0.3 & 0 & 0 \\ 0 & 0 & 0.4 & 0 \\ 0 & 0 & 0 & 0.5 \\ 0 & 0 & 0 & 0 \end{bmatrix}.$$

We need the matrix $(I-N)^{-1}A$, where A is the matrix formed by deleting the rows of T corresponding to the absorbing states, viz. the last two rows, and then forming A from the absorbing columns, viz. the last two columns:

$$A = \begin{bmatrix} 0.7 & 0 \\ 0.6 & 0 \\ 0.5 & 0 \\ 0.4 & 0.6 \end{bmatrix}.$$

Thus

$$\begin{aligned}(I-N)^{-1}A &= \begin{bmatrix} 1 & 0.3 & 0.12 & 0.06 \\ 0 & 1 & 0.4 & 0.2 \\ 0 & 0 & 1 & 0.5 \\ 0 & 0 & 0 & 1 \end{bmatrix} \begin{bmatrix} 0.7 & 0 \\ 0.6 & 0 \\ 0.5 & 0 \\ 0.4 & 0.6 \end{bmatrix} \\ &= \begin{bmatrix} 0.964 & 0.036 \\ 0.88 & 0.12 \\ 0.7 & 0.3 \\ 0.4 & 0.6 \end{bmatrix}\end{aligned}$$

Rows 1 through 4 are the non-absorbing states and columns 1 and 2 are the absorbing states. From row 1, column 1 the probability that a new account will eventually be paid is 96.4%.

5d. From row 2, column 2 in $(I-N)^{-1}A$ the probability that a 1 month overdue account will eventually become a bad debt is 12%.

5e. From row 1 of $(I-N)^{-1}$ the average new customer will take $1+0.3+0.12+0.06=1.48$ months to pay.

5f. \$4500; \$54,000

Review Exercises

1a.

$$\begin{aligned} P &= [0.12 \quad 0.88], \\ T &= \begin{bmatrix} 0.62 & 0.38 \\ 0.16 & 0.84 \end{bmatrix}, \\ PT &= [0.2152 \quad 0.7848] \end{aligned}$$

After three years Toyonda's market share will be 21.52%. (Remember that the survey shows that auto owners buy a new car every three years, *not* every year.)

1b.

$$\begin{aligned} PT^2 &= (PT)T \\ &= [0.2152 \quad 0.7848]\begin{bmatrix} 0.62 & 0.38 \\ 0.16 & 0.84 \end{bmatrix} \\ &= [0.258992 \quad 0.741008] \end{aligned}$$

After 6 years Toyonda's market share will be 25.90%.

1c.

$$\begin{aligned} LT &= L \\ [x \quad y]\begin{bmatrix} 0.62 & 0.38 \\ 0.16 & 0.84 \end{bmatrix} &= [x \quad y] \\ [0.62x + 0.16y \quad 0.38x + 0.84y] &= [x \quad y] \end{aligned}$$

$$\begin{aligned} 0.62x + 0.16y &= x, \\ 0.38x + 0.84y &= y \end{aligned}$$

$$\begin{aligned} -0.38x + 0.16y &= 0, \\ 0.38x - 0.16y &= 0, \\ x + y &= 1 \end{aligned}$$

Using Gauss-Jordan or a calculator on the matrix

$$\begin{bmatrix} -0.38 & 0.16 & 0 \\ 0.38 & -0.16 & 0 \\ 1 & 1 & 1 \end{bmatrix},$$

we obtain

$$\begin{bmatrix} 1 & 0 & 0.296296296296 \\ 0 & 1 & 0.703703703704 \\ 0 & 0 & 0 \end{bmatrix}.$$

Thus

$$L = [x \quad y] = [0.296296296296 \quad 0.703703703704].$$

Toyonda's long range market share is predicted to be 29.63%.

1d. Current trends continue, and the survey accurately reflects current purchases.

2a.

$$P = [0.23 \quad 0.18 \quad 0.59],$$
$$T = \begin{bmatrix} 0.92 & 0.03 & 0.05 \\ 0.01 & 0.88 & 0.11 \\ 0.02 & 0.04 & 0.94 \end{bmatrix},$$
$$\begin{aligned} PT^2 &= (PT)T \\ &= \left([0.23 \quad 0.18 \quad 0.59] \begin{bmatrix} 0.92 & 0.03 & 0.05 \\ 0.01 & 0.88 & 0.11 \\ 0.02 & 0.04 & 0.94 \end{bmatrix} \right) \begin{bmatrix} 0.92 & 0.03 & 0.05 \\ 0.01 & 0.88 & 0.11 \\ 0.02 & 0.04 & 0.94 \end{bmatrix} \\ &= [0.2252 \quad 0.1889 \quad 0.5859] \begin{bmatrix} 0.92 & 0.03 & 0.05 \\ 0.01 & 0.88 & 0.11 \\ 0.02 & 0.04 & 0.94 \end{bmatrix} \\ &= [0.220791 \quad 0.196424 \quad 0.582785] \end{aligned}$$

More condominiums will be needed.

2b.

$$LT = L$$
$$[x \quad y \quad z] \begin{bmatrix} 0.92 & 0.03 & 0.05 \\ 0.01 & 0.88 & 0.11 \\ 0.02 & 0.04 & 0.94 \end{bmatrix} = [x \quad y \quad z]$$
$$[0.92x + 0.01y + 0.02z \quad 0.03x + 0.88y + 0.04z \quad 0.05x + 0.11y + 0.94z] = [x \quad y \quad z]$$

$$\begin{aligned} 0.92x + 0.01y + 0.02z &= x, \\ 0.03x + 0.88y + 0.04z &= y, \\ 0.05x + 0.11y + 0.94z &= z \end{aligned}$$

$$\begin{aligned} -0.8x + 0.01y + 0.02z &= 0, \\ 0.03x - 0.12y + 0.04z &= 0, \\ 0.05x + 0.11y - 0.06z &= 0, \\ x + y + z &= 1 \end{aligned}$$

Note that row 1 $= -$row 2 $-$ row 3, thus row 1 can be discarded. Using Gauss-Jordan or a calculator on the matrix

$$\begin{bmatrix} 0.03 & -0.12 & 0.04 & 0 \\ 0.05 & 0.11 & -0.06 & 0 \\ 1 & 1 & 1 & 1 \end{bmatrix},$$

we obtain

$$\begin{bmatrix} 1 & 0 & 0 & 0.176100628931 \\ 0 & 1 & 0 & 0.238993710692 \\ 0 & 0 & 1 & 0.584905660377 \end{bmatrix}.$$

Thus

$$L = [x \quad y \quad z] = [0.176100628931 \quad 0.238993710692 \quad 0.584905660377].$$

Significantly fewer apartments will be needed, and more condominiums will be needed.

2c. One important ignored factor is that this prediction assumes that this is a closed system; allowances are not made for people moving away from Foxtail County and moving in from outside as well.

3a. We have the non-absorbing states *first-year subscribers*, *second-year subscribers*, and *long-term subscribers*, and an absorbing state *cancel*. Following this order, we have the transition matrix

$$T = \begin{bmatrix} 0 & 0.75 & 0 & 0.25 \\ 0 & 0 & 0.68 & 0.32 \\ 0 & 0 & 0.92 & 0.08 \\ 0 & 0 & 0 & 1 \end{bmatrix}.$$

Recall that N is the matrix which contains only the non-absorbing states:

$$N = \begin{bmatrix} 0 & 0.75 & 0 \\ 0 & 0 & 0.68 \\ 0 & 0 & 0.92 \end{bmatrix}.$$

Now,

$$\begin{aligned}(I-N)^{-1} &= \left(\begin{bmatrix} 1 & 0 & 0 \\ 0 & 1 & 0 \\ 0 & 0 & 1 \end{bmatrix} - \begin{bmatrix} 0 & 0.75 & 0 \\ 0 & 0 & 0.68 \\ 0 & 0 & 0.92 \end{bmatrix} \right)^{-1} \\ &= \begin{bmatrix} 1 & -0.75 & 0 \\ 0 & 1 & -0.68 \\ 0 & 0 & 0.08 \end{bmatrix}^{-1} \\ &= \begin{bmatrix} 1 & 0.75 & 6.375 \\ 0 & 1 & 8.5 \\ 0 & 0 & 12.5 \end{bmatrix}.\end{aligned}$$

From $(I-N)^{-1}$ a new subscriber will be subscribed for $1+0.75+6.375 = 8.125$ yrs.

3b. From $(I-N)^{-1}$ a second year subscriber will be subscribed for $1+8.5 = 9.5$ more years.

4. The non-absorbing states are *10, 20, 30, 40,* and *50,* and the absorbing states are *60* and *broke*, indicating Lucky Larry's cash holdings. Following this order we have the transition matrix

$$T = \begin{bmatrix} 0 & 0.49 & 0 & 0 & 0 & 0 & 0.51 \\ 0.51 & 0 & 0.49 & 0 & 0 & 0 & 0 \\ 0 & 0.51 & 0 & 0.49 & 0 & 0 & 0 \\ 0 & 0 & 0.51 & 0 & 0.49 & 0 & 0 \\ 0 & 0 & 0 & 0.51 & 0 & 0.49 & 0 \\ 0 & 0 & 0 & 0 & 0 & 1 & 0 \\ 0 & 0 & 0 & 0 & 0 & 0 & 1 \end{bmatrix}.$$

Recall that N is the matrix which contains only the non-absorbing states:

$$N = \begin{bmatrix} 0 & 0.49 & 0 & 0 & 0 \\ 0.51 & 0 & 0.49 & 0 & 0 \\ 0 & 0.51 & 0 & 0.49 & 0 \\ 0 & 0 & 0.51 & 0 & 0.49 \\ 0 & 0 & 0 & 0.51 & 0 \end{bmatrix}.$$

We need the matrix $(I-N)^{-1}A$, where A is the matrix formed by deleting the rows of T corresponding to the absorbing states, viz. the last two rows, and then forming A from the absorbing columns, viz. the last two columns:

$$A = \begin{bmatrix} 0 & 0.51 \\ 0 & 0 \\ 0 & 0 \\ 0 & 0 \\ 0.49 & 0 \end{bmatrix}.$$

Thus

$$\begin{aligned}(I-N)^{-1}A &\approx \begin{bmatrix} 1.6658 & 1.3054 & 0.9592 & 0.6266 & 0.3070 \\ 1.3587 & 2.6642 & 1.9577 & 1.2788 & 0.6266 \\ 1.0392 & 2.0376 & 2.9968 & 1.9577 & 0.9592 \\ 0.7065 & 1.3854 & 2.0376 & 2.6642 & 1.3054 \\ 0.3603 & 0.7065 & 1.0392 & 1.3587 & 1.6658 \end{bmatrix} \begin{bmatrix} 0 & 0.51 \\ 0 & 0 \\ 0 & 0 \\ 0 & 0 \\ 0.49 & 0 \end{bmatrix} \\ &\approx \begin{bmatrix} 0.1505 & 0.8495 \\ 0.3070 & 0.6930 \\ 0.4700 & 0.5300 \\ 0.6397 & 0.3603 \\ 0.8162 & 0.1838 \end{bmatrix}\end{aligned}$$

Rows 1 through 5 are the non-absorbing states and columns 1 and 2 are the absorbing states. From row 3, column 1 the probability that he makes enough to go home is 47%.

Chapter 8 Game Theory

Section 8.1 Introduction to Game Theory

1. To determine whether the zero-sum game is strictly determined, we must find an entry in the reward matrix which is at once the largest of the row minimums and the smallest of the column maximums. In this case the largest of the row minimums is 2 and the smallest of the column maximums is also 2; thus this is strictly determined. (The saddle point is 2.)

The row player should consistently choose row 1 (the row containing the saddle point), and the column player should consistently choose column 1 (the column containing the saddle point).

The value of the game is the value of the saddle point, viz. 2.

The game favors the row player.

5. To determine whether the zero-sum game is strictly determined, we must find an entry in the reward matrix which is at once the largest of the row minimums and the smallest of the column maximums. In this case the largest of the row minimums is -1 and the smallest of the column maximums is 3; thus this is not strictly determined.

The row player's highest reward is 5 in row 1, so *row 1* is played first. Since row 1 is being played, the column player's highest reward (in row 1) is -2 in column 2, so *column 2* is then played. In column 2 the row player's highest reward is 3 in row 2, so *row 2* is then played. In row 2 the column player's highest reward is -6 in column 1, so *column 1* is played.

9. To determine whether the zero-sum game is strictly determined, we must find an entry in the reward matrix which is at once the largest of the row minimums and the smallest of the column maximums. In this case the largest of the row minimums is 0 and the smallest of the column maximums is also 0; thus this is strictly determined. (The saddle point is 0.)

The row player should consistently choose row 2 (the row containing the saddle point), and the column player should consistently choose column 1 (the column containing the saddle point).

The value of the game is the value of the saddle point, viz. 0.

The game is fair.

13a. First we make a reward matrix with the given percentages in which RunzGood is the row player and Petrolia is the column player:

$$\begin{bmatrix} 3 & -2 & -5 \\ 1 & 0 & -3 \\ 4 & 1 & 2 \end{bmatrix}$$

This is a zero-sum game since if one station gains n% of the two stations' combined business, then the other station loses n% of the two stations' combined business.

13b. To determine whether the zero-sum game is strictly determined, we must find an entry in the reward matrix which is at once the largest of the row minimums and the smallest of the column maximums. In this case the largest of the row minimums is 1 (in row 3) and the smallest of the column maximums is also 1 (in column 2); thus this is strictly determined. (The saddle point is 1 in row 3, column 2.)

13c. The row player (RunzGood) should consistently choose row 3 (the row containing the saddle point), i.e., it should always lower prices, and the column player (Petrolia) should consistently choose column 2 (the column containing the saddle point), i.e., it should not change prices.

The value of the game is the value of the saddle point, viz. 1.

The game favors RunzGood; it will continue to gain 1% of the stations' combined business.

17a. First we make a reward matrix with the given percentages in which Brown is the row player and Jordan is the column player:

$$\begin{bmatrix} 3 & 5 & 0 \\ -4 & 0 & 2 \\ 0 & -6 & -3 \end{bmatrix}$$

This is a zero-sum game since if one candidate gains n% of the votes in the other half of the state, then the other candidate loses n% of the votes in "his" half of the state.

17b. To determine whether the zero-sum game is strictly determined, we must find an entry in the reward matrix which is at once the largest of the row minimums and the smallest of the column maximums. In this case the largest of the row minimums is 0 (in row 1) and the smallest of the column maximums is 2 (in column 3); thus this is not strictly determined.

17c. First, Brown would emphasize the north since his largest reward (5%) is in row 1; then Jordan would emphasize the south since this is the best he can do in row 1 (no net change in row 1, column 3); then Brown would split the money equally since his best reward in column 3 is 2 (in row 2); then Jordan would emphasixe the north since his best reward in row 2 is a gain of 4% (in column 1).

21. Their losses are kept to a minimum by following the best pure strategy.

1a. Let the odd player be the row player. The reward matrix is

$$\begin{bmatrix} -1 & 1 \\ 1 & -1 \end{bmatrix}.$$

To determine whether the zero-sum game is strictly determined, we must find an entry in the reward matrix which is at once the largest of the row minimums and the smallest of the column maximums. In this case the largest of the row minimums is -1 (in both rows) and the smallest of the column maximums is 1 (in both columns); thus this is not strictly determined.

1b.

$$\begin{aligned} E(pure) &= P_r \cdot R \\ &= \begin{bmatrix} \frac{1}{4} & \frac{3}{4} \end{bmatrix} \begin{bmatrix} -1 & 1 \\ 1 & -1 \end{bmatrix} \\ &= \begin{bmatrix} \frac{1}{2} & -\frac{1}{2} \end{bmatrix} \end{aligned}$$

The expected value is the smaller entry in $E(pure)$, viz., $-\frac{1}{2}$. Thus the game favors the even player if the odd player adopts a random strategy and the even player adopts a pure strategy.

1c. Let p_1 be the probability with which the odd player shows one finger; then $1 - p_1$ is the probability with which the odd player shows two fingers. Then

$$\begin{aligned} E(pure) &= P_r \cdot R \\ &= \begin{bmatrix} p_1 & 1 - p_1 \end{bmatrix} \begin{bmatrix} -1 & 1 \\ 1 & -1 \end{bmatrix} \\ &= \begin{bmatrix} -p_1 + 1 - p_1 & p_1 + p_1 - 1 \end{bmatrix} \\ &= \begin{bmatrix} 1 - 2p_1 & 2p_1 - 1 \end{bmatrix} \end{aligned}$$

The expected value is $-2p_1 + 1$ if the even player always responds with one finger, and it is $2p_1 - 1$ if the even player always responds with two fingers. These are graphed as follows, with p_1 on the horizontal axis and the expected values on the vertical axis:

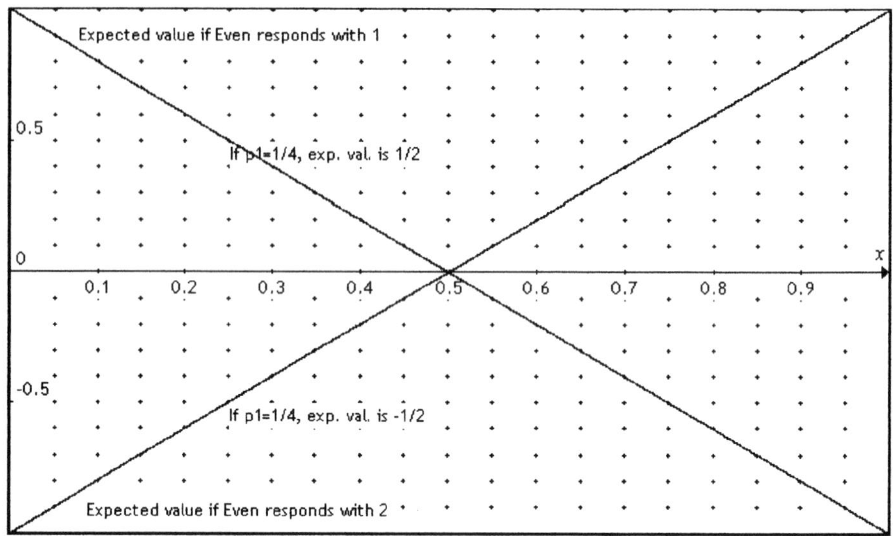

1d.

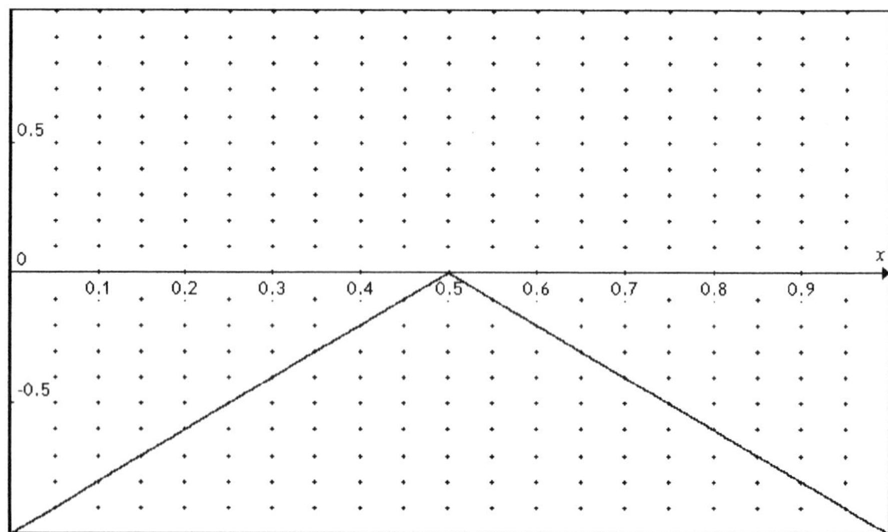

The even player will always respond to the odd player such that Odd's winnings are minimized; thus the higher expected values, as shown in the top portion of the graph in 1c, are not attainable (assuming that Even is a capable player).

1e. The game's highest expected value is found at the intersection of the two lines in 1d. This is the point where

$$
\begin{aligned}
-2p_1 + 1 &= 2p_1 - 1 \\
2 &= 4p_1 \\
p_1 &= \frac{1}{2}
\end{aligned}
$$

The expected value of the game is then $2(\frac{1}{2}) - 1 = 0$.

1f. Odd's optimal randomized strategy is to show one finger half the time and two fingers half the time.

1g. If Odd follows the optimal randomized strategy, the game is fair since its expected value is then zero.

5a. To determine whether the zero-sum game is strictly determined, we must find an entry in the reward matrix which is at once the largest of the row minimums and the smallest of the column maximums. In this case the largest of the row minimums is 3 (in row 2) and the smallest of the column maximums is 3 (in column 2); thus this is strictly determined.

5b. N/A

5c.

$$
\begin{aligned}
[\,p_1 \quad 1-p_1\,]\begin{bmatrix} 5 & 2 \\ 7 & 3 \end{bmatrix} &= [\,5p_1 + 7 - 7p_1 \quad 2p_1 + 3 - 3p_1\,] \\
&= [\,-2p_1 + 7 \quad -p_1 + 3\,]
\end{aligned}
$$

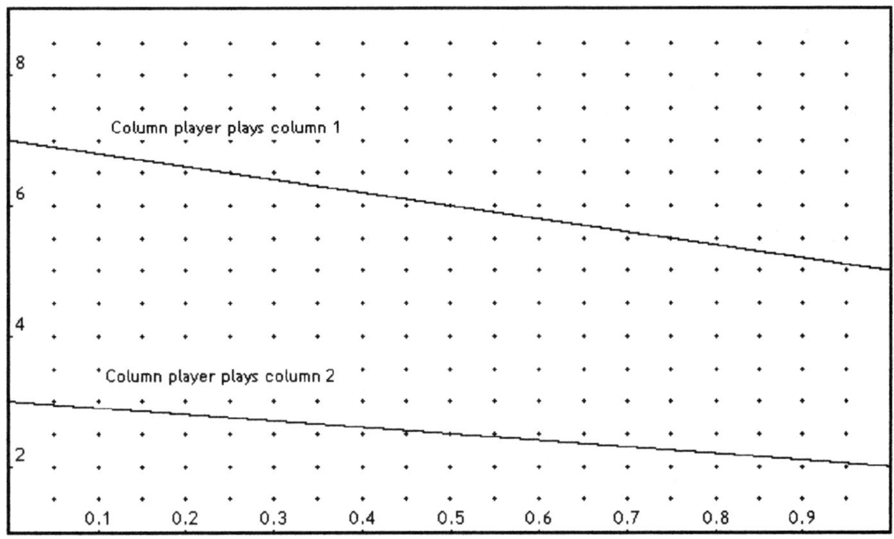

From the graph we see that the game always favors the row player (since the expected value is always above the horizontal axis), but the column player can minimize his losses by always playing column 2. (This is to be expected, since it is strictly determined.) The row player should play row 1 none of the time ($p_1 = 0$ maximizes the row player's winnings) and row 2 all of the time. The expected value is $-(0) + 3 = 3$. Could the column player do better with a randomized strategy?

$$\begin{aligned}[p_1 \quad 1-p_1]\begin{bmatrix}5 & 2\\ 7 & 3\end{bmatrix}\begin{bmatrix}q_1\\ 1-q_1\end{bmatrix} &= [-2p_1+7 \quad -p_1+3]\begin{bmatrix}q_1\\ 1-q_1\end{bmatrix}\\ &= [(-2p_1+7)q_1+(-p_1+3)(1-q_1)]\\ &= [-2p_1q_1+7q_1-p_1+p_1q_1+3-3q_1]\\ &= [-2(0)q_1+7q_1-(0)+(0)q_1+3-3q_1]\\ &= [7q_1+3-3q_1]\\ &= [4q_1+3]\end{aligned}$$

The expected value is $4q_1+3$; that is, the column player must always choose to play column 2 (which means that $q_1=0$) in order to minimize the expected value; if the column player does this the expected value will always be $4(0)+3=3$, but if the column player ever plays column 1, then $q_1>0$ and the expected value will be greater than 3.

9a. To determine whether the zero-sum game is strictly determined, we must find an entry in the reward matrix which is at once the largest of the row minimums and the smallest of the column maximums. In this case the largest of the row minimums is 4 (in row 1) and the smallest of the column maximums is 4 (in column 2); thus this is strictly determined. (The saddle point is 4.)

9b. The row player should consistently choose row 1 (the row containing the saddle point), and the column player should consistently choose column 2 (the column containing the saddle point).

The value of the game is the value of the saddle point, viz. 4.

The game favors the row player.

9c. N/A

13a. Normal and dry conditions, in order of preference, are better for wheat; wet and normal conditions, in order of preference, are better for rice.

13b. We can make a reward matrix with row 1 corresponding to rice, row 2 corresponding to wheat, column 1 corresponding to a wet year, column 2 corresponding to a normal year, and column 3 corresponding to a dry year (entries are millions of dollars):

$$\begin{bmatrix}10 & 5 & -5\\ -3 & 12 & 3\end{bmatrix}.$$

Now, using a reduced reward matrix (column 2 is a dominated "strategy"), we have

$$\begin{aligned}[p_1 \quad 1-p_1]\begin{bmatrix}10 & -5\\ -3 & 3\end{bmatrix} &= [10p_1-3+3p_1 \quad -5p_1+3-3p_1]\\ &= [13p_1-3 \quad -8p_1+3].\end{aligned}$$

Set the entries equal to each other and solve for p_1:

$$\begin{aligned} 13p_1 - 3 &= -8p_1 + 3 \\ 13p_1 + 8p_1 &= 3 + 3 \\ 21p_1 &= 6 \\ p_1 &= \frac{6}{21} \\ &= \frac{2}{7} \end{aligned}$$

Thus $\frac{2}{7}$ of the tract should be planted with rice and $1 - \frac{2}{7} = \frac{5}{7}$ of the tract should be planted with wheat.

13c. Wet: $\frac{2}{7}(10) - \frac{5}{7}(3) = \frac{5}{7} \approx 0.714$, or a profit of about $714,000 will be made. Normal: $\frac{2}{7}(5) + \frac{5}{7}(12) = 10$, or a profit of $10,000,000 will be made. Dry: $\frac{2}{7}(-5) + \frac{5}{7}(3) = \frac{5}{7}$, or a profit of about $714,000 will be made.

Section 8.3 Game Theory and Linear Programming

1. To determine whether the game is strictly determined, we must find an entry in the reward matrix which is at once the largest of the row minimums and the smallest of the column maximums. In this case the largest of the row minimums is -8 (in row 1) and the smallest of the column maximums is 9 (in column 2); thus this is not strictly determined.

We need to change the reward matrix R to R' which will have all positive entries (yet entries which are as small as possible to simplify computations). Thus we add 12 to each entry in R (equivalently, add a constant matrix with entries of 12 to R) to obtain R':

$$\begin{aligned} R' &= R + \begin{bmatrix} 12 & 12 \\ 12 & 12 \end{bmatrix} \\ &= \begin{bmatrix} 10 & -8 \\ -11 & 9 \end{bmatrix} + \begin{bmatrix} 12 & 12 \\ 12 & 12 \end{bmatrix} \\ &= \begin{bmatrix} 10+12 & -8+12 \\ -11+12 & 9+12 \end{bmatrix} \\ &= \begin{bmatrix} 22 & 4 \\ 1 & 21 \end{bmatrix} \end{aligned}$$

Let p_1 be the probability with which the row player plays row 1; then $p_2 = 1 - p_1$ is the probability that the row player plays row 2, and the expected values of the column player's pure strategies are found in the matrix

$$\begin{aligned} E(pure) &= P_r \cdot R' \\ &= \begin{bmatrix} p_1 & p_2 \end{bmatrix} \begin{bmatrix} 22 & 4 \\ 1 & 21 \end{bmatrix} \\ &= \begin{bmatrix} 22p_1 + p_2 & 4p_1 + 21p_2 \end{bmatrix} \\ &= \begin{bmatrix} 22p_1 + (1-p_1) & 4p_1 + 21(1-p_1) \end{bmatrix} \\ &= \begin{bmatrix} 21p_1 + 1 & 4p_1 + 21 - 21p_1 \end{bmatrix} \\ &= \begin{bmatrix} 21p_1 + 1 & -17p_1 + 21 \end{bmatrix}. \end{aligned}$$

Since the column player wants to minimize the expected value, he or she will choose the smaller of $21p_1 + 1$ and $-17p_1 + 21$. Thus $E' \leq 21p_1 + 1$ and $E' \leq -17p_1 + 21$, where $E' = E + 12$, E is the expected value of the game associated with the original reward matrix R, and E' is the expected value of the game associated with the reward matrix R'. The objective of the linear programming problem is to maximize E' since the row player wants to maximize his or her reward. Constraints with slack variables s_n ($n = 1, 2, 3$) and objective function:

$$\begin{aligned} &C_1 : E' - 21p_1 + s_1 = 1, \\ &C_2 : E' + 17p_1 + s_2 = 21, \\ &C_3 : p_1 + s_3 = 1, \\ &\text{Obj. func.}: -E' + z = 0. \end{aligned}$$

If the (ascending) columns correspond to the order $E', p_1, s_1, s_2, s_3, z,$ and the constants, we have the first simplex matrix

$$\begin{bmatrix} 1 & -21 & 1 & 0 & 0 & 0 & 1 \\ 1 & 17 & 0 & 1 & 0 & 0 & 21 \\ 0 & 1 & 0 & 0 & 1 & 0 & 1 \\ -1 & 0 & 0 & 0 & 0 & 1 & 0 \end{bmatrix}.$$

Recall that we first select the most negative entry in the objective function (last) row; this marks the pivot column. Then we divide the last entry in each constraint row (first three rows) by the corresponding entry in the pivot column. The smallest nonnegative quotient marks the pivot row. Once we find the pivot entry, we proceed with the Guass-Jordan method.

We repeat the above procedure until we arrive at a matrix with no negative entries in the last row; this is the final simplex matrix.

The first pivot entry is in row 1, column 1 (above matrix). The procedure yields:

$$\begin{bmatrix} 1 & -21 & 1 & 0 & 0 & 0 & 1 \\ 0 & 38 & -1 & 1 & 0 & 0 & 20 \\ 0 & 1 & 0 & 0 & 1 & 0 & 1 \\ 0 & -21 & 1 & 0 & 0 & 1 & 1 \end{bmatrix}.$$

The next pivot entry is in row 2, column 2. The procedure yields the final simplex matrix:

$$\begin{bmatrix} 1 & 0 & 0.4474 & 0.5526 & 0 & 0 & 12.0526 \\ 0 & 1 & -0.0263 & 0.0263 & 0 & 0 & 0.5263 \\ 0 & 0 & 0.0263 & -0.0263 & 1 & 0 & 0.4737 \\ 0 & 0 & 0.4474 & 0.5526 & 0 & 1 & 12.0526 \end{bmatrix}.$$

Thus, recalling the column order $E', p_1, s_1, s_2, s_3, z,$ and the constants, we have $E' = 12.0526$, $p_1 = 0.5263$, and $p_2 = 1 - p_1 = s_3 = 0.4737$. But $E' = E + 12$, whence $E = E' - 12 = 12.0526 - 12 = 0.0526$.

5. To determine whether the game is strictly determined, we must find an entry in the reward matrix which is at once the largest of the row minimums and the smallest of the column maximums. In this case the largest of the row minimums is -2 (in rows 1 and 2) and the smallest of the column maximums is 1 (in columns 1 and 4); thus this is not strictly determined.

Dominated strategies which can be eliminated are row 3 and column 4.

We need to change the (modified) reward matrix R to R' which will have all positive entries (yet entries which are as small as possible to simplify computations). Thus we add 3 to each entry in R (equivalently, add a constant matrix with entries of 3 to R) to obtain R':

$$\begin{aligned} R' &= R + \begin{bmatrix} 3 & 3 & 3 \\ 3 & 3 & 3 \end{bmatrix} \\ &= \begin{bmatrix} 1 & 2 & -2 \\ -1 & -2 & 3 \end{bmatrix} + \begin{bmatrix} 3 & 3 & 3 \\ 3 & 3 & 3 \end{bmatrix} \\ &= \begin{bmatrix} 1+3 & 2+3 & -2+3 \\ -1+3 & -2+3 & 3+3 \end{bmatrix} \\ &= \begin{bmatrix} 4 & 5 & 1 \\ 2 & 1 & 6 \end{bmatrix} \end{aligned}$$

Let p_1 be the probability with which the row player plays row 1; then $p_2 = 1 - p_1$ is the probability that the row player plays row 2, and the expected values of the column player's pure strategies are found in the matrix

$$\begin{aligned}E(pure) &= P_r \cdot R' \\ &= [\,p_1 \quad p_2\,]\begin{bmatrix}4 & 5 & 1\\ 2 & 1 & 6\end{bmatrix} \\ &= [\,4p_1 + 2p_2 \quad 5p_1 + p_2 \quad p_1 + 6p_2\,] \\ &= [\,4p_1 + 2(1-p_1) \quad 5p_1 + (1-p_1) \quad p_1 + 6(1-p_1)\,] \\ &= [\,4p_1 + 2 - 2p_1 \quad 4p_1 + 1 \quad p_1 + 6 - 6p_1\,] \\ &= [\,2p_1 + 2 \quad 4p_1 + 1 \quad -5p_1 + 6\,].\end{aligned}$$

Since the column player wants to minimize the expected value, he or she will choose the smallest of $2p_1 + 2$, $4p_1 + 1$, and $-5p_1 + 6$. Thus $E' \le 2p_1 + 2$, $E' \le 4p_1 + 1$, and $E' \le -5p_1 + 6$, where $E' = E + 3$, E is the expected value of the game associated with the original reward matrix R, and E' is the expected value of the game associated with the reward matrix R'. The objective of the linear programming problem is to maximize E' since the row player wants to maximize his or her reward. Constraints with slack variables s_n $(n = 1, 2, 3, 4)$ and objective function:

$$\begin{aligned}&C_1 : E' - 2p_1 + s_1 = 2,\\ &C_2 : E' - 4p_1 + s_2 = 1,\\ &C_3 : E' + 5p_1 + s_3 = 6,\\ &C_4 : p_1 + s_4 = 1,\\ &\text{Obj. func.}: -E' + z = 0.\end{aligned}$$

If the (ascending) columns correspond to the order $E', p_1, s_1, s_2, s_3, s_4, z,$ and the constants, we have the first simplex matrix

$$\begin{bmatrix}1 & -2 & 1 & 0 & 0 & 0 & 0 & 2\\ 1 & -4 & 0 & 1 & 0 & 0 & 0 & 1\\ 1 & 5 & 0 & 0 & 1 & 0 & 0 & 6\\ 0 & 1 & 0 & 0 & 0 & 1 & 0 & 1\\ -1 & 0 & 0 & 0 & 0 & 0 & 1 & 0\end{bmatrix}.$$

Recall that we first select the most negative entry in the objective function (last) row; this marks the pivot column. Then we divide the last entry in each constraint row (first four rows) by the corresponding entry in the pivot column. The smallest nonnegative quotient marks the pivot row. Once we find the pivot entry, we proceed with the Guass-Jordan method.

We repeat the above procedure until we arrive at a matrix with no negative entries in the last row; this is the final simplex matrix.

The first pivot entry is in row 2, column 1 (above matrix). The procedure yields:

$$\begin{bmatrix}0 & 2 & 1 & -1 & 0 & 0 & 0 & 1\\ 1 & -4 & 0 & 1 & 0 & 0 & 0 & 1\\ 0 & 9 & 0 & -1 & 1 & 0 & 0 & 5\\ 0 & 1 & 0 & 0 & 0 & 1 & 0 & 1\\ 0 & -4 & 0 & 1 & 0 & 0 & 1 & 1\end{bmatrix}.$$

The next pivot entry is in row 1, column 2. The procedure yields:

$$\begin{bmatrix}0 & 1 & 0.5 & -0.5 & 0 & 0 & 0 & 0.5\\ 1 & 0 & 2 & -1 & 0 & 0 & 0 & 3\\ 0 & 0 & -4.5 & 3.5 & 1 & 0 & 0 & 0.5\\ 0 & 0 & -0.5 & 0.5 & 0 & 1 & 0 & 0.5\\ 0 & 0 & 2 & -1 & 0 & 0 & 1 & 3\end{bmatrix}.$$

The next pivot entry is in row 1, column 2. The procedure yields the final simplex matrix:

$$\begin{bmatrix} 0 & 1 & -0.1429 & 0 & 0.1429 & 0 & 0 & 0.5714 \\ 1 & 0 & 0.7143 & 0 & 0.2857 & 0 & 0 & 3.1429 \\ 0 & 0 & -1.2857 & 1 & 0.2857 & 0 & 0 & 0.1429 \\ 0 & 0 & 0.1429 & 0 & -0.1429 & 1 & 0 & 0.4286 \\ 0 & 0 & 0.7143 & 0 & 0.2857 & 0 & 1 & 3.1429 \end{bmatrix}.$$

Thus, recalling the column order $E', p_1, s_1, s_2, s_3, s_4, z,$ and the constants, we have $p_1 = 0.5714$, and $p_2 = 1 - p_1 = s_4 = 0.4286$. These are the probabilities for row 1 and row 2, respectively; row 3 should never be chosen since it's a dominated strategy.

9. To determine whether the game is strictly determined, we must find an entry in the reward matrix which is at once the largest of the row minimums and the smallest of the column maximums. In this case the largest of the row minimums is 0 (in row 1) and the smallest of the column maximums is 2 (in column 3); thus this is not strictly determined.

Dominated strategies which can be eliminated are row 3 and column 2.

We need to change the reward matrix R to R' which will have all positive entries (yet entries which are as small as possible to simplify computations). Thus we add 5 to each entry in R (equivalently, add a constant matrix with entries of 5 to R) to obtain R':

$$\begin{aligned} R' &= R + \begin{bmatrix} 5 & 5 \\ 5 & 5 \end{bmatrix} \\ &= \begin{bmatrix} 3 & 0 \\ -4 & 2 \end{bmatrix} + \begin{bmatrix} 5 & 5 \\ 5 & 5 \end{bmatrix} \\ &= \begin{bmatrix} 3+5 & 0+5 \\ -4+5 & 2+5 \end{bmatrix} \\ &= \begin{bmatrix} 8 & 5 \\ 1 & 7 \end{bmatrix} \end{aligned}$$

Let p_1 be the probability with which Brown emphasizes the north; then $p_2 = 1 - p_1$ is the probability that Brown splits the money equally, and the expected values of Jordan's pure strategies are found in the matrix

$$\begin{aligned} E(pure) &= P_r \cdot R' \\ &= [\,p_1 \quad p_2\,]\begin{bmatrix} 8 & 5 \\ 1 & 7 \end{bmatrix} \\ &= [\,8p_1 + p_2 \quad 5p_1 + 7p_2\,] \\ &= [\,8p_1 + (1 - p_1) \quad 5p_1 + 7(1 - p_1)\,] \\ &= [\,7p_1 + 1 \quad 5p_1 + 7 - 7p_1\,] \\ &= [\,7p_1 + 1 \quad -2p_1 + 7\,]. \end{aligned}$$

Since Jordan wants to minimize the expected value, he will choose the smaller of $7p_1 + 1$ and $-2p_1 + 7$. Thus $E' \le 7p_1 + 1$ and $E' \le -2p_1 + 7$, where $E' = E + 5$, E is the expected value of the game associated with the original reward matrix R, and E' is the expected value of the game associated with the reward matrix R'. The objective of the linear programming problem is to maximize E' since Brown wants to maximize his reward. Constraints with slack variables s_n ($n = 1, 2, 3$) and objective function:

$$\begin{aligned}&C_1 : E' - 7p_1 + s_1 = 1,\\ &C_2 : E' + 2p_1 + s_2 = 7,\\ &C_3 : p_1 + s_3 = 1,\\ &\text{Obj. func.: } -E' + z = 0.\end{aligned}$$

If the (ascending) columns correspond to the order $E', p_1, s_1, s_2, s_3, z,$ and the constants, we have the first simplex matrix

$$\begin{bmatrix} 1 & -7 & 1 & 0 & 0 & 0 & 1 \\ 1 & 2 & 0 & 1 & 0 & 0 & 7 \\ 0 & 1 & 0 & 0 & 1 & 0 & 1 \\ -1 & 0 & 0 & 0 & 0 & 1 & 0 \end{bmatrix}.$$

Recall that we first select the most negative entry in the objective function (last) row; this marks the pivot column. Then we divide the last entry in each constraint row (first three rows) by the corresponding entry in the pivot column. The smallest nonnegative quotient marks the pivot row. Once we find the pivot entry, we proceed with the Guass-Jordan method.

We repeat the above procedure until we arrive at a matrix with no negative entries in the last row; this is the final simplex matrix.

The first pivot entry is in row 1, column 1 (above matrix). The procedure yields:

$$\begin{bmatrix} 1 & -7 & 1 & 0 & 0 & 0 & 1 \\ 0 & 9 & -1 & 1 & 0 & 0 & 6 \\ 0 & 1 & 0 & 0 & 1 & 0 & 1 \\ 0 & -7 & 1 & 0 & 0 & 1 & 1 \end{bmatrix}.$$

The next pivot entry is in row 2, column 2. The procedure yields the final simplex matrix:

$$\begin{bmatrix} 1 & 0 & 0.22 & 0.78 & 0 & 0 & 5.67 \\ 0 & 1 & -0.11 & 0.11 & 0 & 0 & 0.67 \\ 0 & 0 & 0.11 & -0.11 & 1 & 0 & 0.33 \\ 0 & 0 & 0.22 & 0.78 & 0 & 1 & 5.67 \end{bmatrix}.$$

Thus, recalling the column order $E', p_1, s_1, s_2, s_3, z,$ and the constants, we have $p_1 = \frac{2}{3}$ and $p_2 = 1 - p_1 = s_3 = \frac{1}{3}$. That is, Brown will emphasize the north 0.67 of the time and will split the money equally 0.33 of the time; he will never emphasize the south.

13. Let Global Fon be the row player. There are three choices for each player: to locate in Backwater; to locate in Podunk; to locate in River City. Thus we obtain a 3×3 reward matrix, with columns and rows corresponding to the order given immediately above.

To obtain the entries for row 1, note that the first entry corresponds to both stores locating in Backwater. In this case the row player will get 50% of the business for the entire tri-city area; hence 50 is entered in row 1, column 1.

Row 1, column 2 corresponds to Global Fon locating in Backwater and U-Call locating in Podunk. This means that Global Fon will get all of Backwater's business, or 20% of the tri-city residents, none of Podunk's business, and half of River City's business, or 25% of the tri-city population, for a total of $20\% + 25\% = 45\%$ of the tri-city population. Thus 45 is entered in row 1, column 2.

Row 1, column 3 corresponds to Global Fon locating in Backwater and U-Call locating in River City. This means that Global Fon will get all of Backwater's business, or 20% of the tri-city residents, none of River City's business, and half of Podunk's business, or 15% of the tri-city population, for a total of $20\% + 15\% = 35\%$ of the tri-city population. Thus 35 is entered in row 1, column 3.

The reward matrix is filled in a similar fashion for the rest of the entries, and we obtain

$$R = \begin{bmatrix} 50 & 45 & 35 \\ 55 & 50 & 45 \\ 65 & 75 & 50 \end{bmatrix}.$$

(Note that the saddle point is in row 3, column 3.)

Now,

$$\begin{aligned} E(pure) &= [\,p_1 \quad p_2 \quad p_3\,] \begin{bmatrix} 50 & 45 & 35 \\ 55 & 50 & 45 \\ 65 & 75 & 50 \end{bmatrix} \\ &= [\,50p_1 + 55p_2 + 65p_3 \quad 45p_1 + 50p_2 + 75p_3 \quad 35p_1 + 45p_2 + 50p_3\,] \\ &= [\,50p_1 + 55p_2 + 65(1 - p_1 - p_2) \quad 45p_1 + 50p_2 + 75(1 - p_1 - p_2) \quad 35p_1 + 45p_2 + 50(1 - p_1 - p_2)\,] \\ &= [\,50p_1 + 55p_2 + 65 - 65p_1 - 65p_2 \quad 45p_1 + 50p_2 + 75 - 75p_1 - 75p_2 \quad 35p_1 + 45p_2 + 50 - 50p_1 - 50p_2\,] \\ &= [\,-15p_1 - 10p_2 + 65 \quad -30p_1 - 25p_2 + 75 \quad -15p_1 - 5p_2 + 50\,]. \end{aligned}$$

Therefore our constraints and objective function are

$$\begin{aligned} E + 15p_1 + 10p_2 + s_1 &= 65, \\ E + 30p_1 + 25p_2 + s_2 &= 75, \\ E + 15p_1 + 5p_2 + s_3 &= 50, \\ p_1 + s_4 &= 1, \\ p_2 + s_5 &= 1, \\ -E + z &= 0. \end{aligned}$$

We obtain the first simplex matrix

$$\begin{bmatrix} 1 & 15 & 10 & 1 & 0 & 0 & 0 & 0 & 0 & 65 \\ 1 & 30 & 25 & 0 & 1 & 0 & 0 & 0 & 0 & 75 \\ 1 & 15 & 5 & 0 & 0 & 1 & 0 & 0 & 0 & 50 \\ 0 & 1 & 0 & 0 & 0 & 0 & 1 & 0 & 0 & 1 \\ 0 & 0 & 1 & 0 & 0 & 0 & 0 & 1 & 0 & 1 \\ -1 & 0 & 0 & 0 & 0 & 0 & 0 & 0 & 1 & 0 \end{bmatrix}.$$

Pivot on the entry in row 3, column 1 to obtain the final simplex matrix:

$$\begin{bmatrix} 0 & 0 & 5 & 1 & 0 & -1 & 0 & 0 & 0 & 15 \\ 0 & 15 & 20 & 0 & 1 & -1 & 0 & 0 & 0 & 25 \\ 1 & 15 & 5 & 0 & 0 & 1 & 0 & 0 & 0 & 50 \\ 0 & 1 & 0 & 0 & 0 & 0 & 1 & 0 & 0 & 1 \\ 0 & 0 & 1 & 0 & 0 & 0 & 0 & 1 & 0 & 1 \\ 0 & 15 & 5 & 0 & 0 & 1 & 0 & 0 & 1 & 50 \end{bmatrix}.$$

We observe that $p_1 = p_2 = 0$, whence $p_3 = 1$ since $p_1 + p_2 + p_3 = 1$. Thus Global Fon should locate in River City; this agrees with the location of the saddle point.

Review Exercises

1a. Let Discs-R-Us be the row player. There are three choices for each player: to locate in Smallville; to locate in Gotham; to locate in Elmwood. Thus we obtain a 3×3 reward matrix, with columns and rows corresponding to the order given immediately above.

To obtain the entries for row 1, note that the first entry corresponds to both stores locating in Smallville. In this case the row player will get 50% of the business for the entire area; hence 50 is entered in row 1, column 1.

Row 1, column 2 corresponds to Discs-R-Us locating in Smallville and 500 Channels! locating in Gotham. This means that Discs-R-Us will get 90% of Smallville's business, or $0.9(0.2) = 18\%$ of the area residents, 10% of Gotham's business, or $0.1(0.35) = 3.5\%$ of the area residents, and half of Elmwood's business, or $45\%/2 = 22.5\%$ of the area's residents, for a total of $18\% + 3.5\% + 22.5\% = 44\%$ of the area's residents. Thus 44 is entered in row 1, column 2.

Row 1, column 3 corresponds to Discs-R-Us locating in Smallville and 500 Channels! locating in Elmwood. This means that Discs-R-Us will get 90% of Smallville's business, or $0.9(0.2) = 18\%$ of the area residents, 10% of Elmwood's business, or $0.1(0.45) = 4.5\%$ of the area residents, and half of Gotham's business, or $35\%/2 = 17.5\%$ of the area's residents, for a total of $18\% + 4.5\% + 17.5\% = 40\%$ of the area's residents. Thus 40 is entered in row 1, column 3.

The reward matrix is filled in a similar fashion for the rest of the entries, and we obtain

$$R = \begin{bmatrix} 50 & 44 & 40 \\ 56 & 50 & 46 \\ 60 & 54 & 50 \end{bmatrix}.$$

To determine whether the game is strictly determined, we must find an entry in the reward matrix which is at once the largest of the row minimums and the smallest of the column maximums. In this case the largest of the row minimums is 50 (in row 3) and the smallest of the column maximums is 50 (in column 3); thus this is strictly determined. The saddle point is 50 in row 3, column 3. Discs-R-Us should locate in Elmwood and 500 Channels! should locate in Elmwood. The value of this fair game is 50%. Each company will get 50% of the central valley area business.

2b. In this case, using the same parameters as above,

$$R = \begin{bmatrix} 60 & 44 & 40 \\ 56 & 60 & 46 \\ 60 & 54 & 60 \end{bmatrix}.$$

To determine whether the game is strictly determined, we must find an entry in the reward matrix which is at once the largest of the row minimums and the smallest of the column maximums. In this case the largest of the row minimums is 54 (in row 3) and the smallest of the column maximums is 60 (in every column); thus this is not strictly determined.

Note that row 1 and column 1 in R are dominated strategies which may be eliminated from consideration for the time being. Hence we obtain a reduced reward matrix $\begin{bmatrix} 60 & 46 \\ 54 & 60 \end{bmatrix}$. Let p_1 be the probability with which Discs-R-Us locates in Gotham; then $1 - p_1$ is the probability with which Discs-R-Us locates in Elmwood. Then

$$\begin{aligned} E(pure) &= P_r \cdot R \\ &= \begin{bmatrix} p_1 & 1-p_1 \end{bmatrix} \begin{bmatrix} 60 & 46 \\ 54 & 60 \end{bmatrix} \\ &= \begin{bmatrix} 60p_1 + 54(1-p_1) & 46p_1 + 60(1-p_1) \end{bmatrix} \\ &= \begin{bmatrix} 60p_1 + 54 - 54p_1 & 46p_1 + 60 - 60p_1 \end{bmatrix} \\ &= \begin{bmatrix} 6p_1 + 54 & -14p_1 + 60 \end{bmatrix} \end{aligned}$$

The expected value is $6p_1 + 54$ if 500 Channels! always responds by locating in Gotham, and it is $-14p_1 + 60$ if 500 Channels! always responds by locating in Elmwood. These are graphed as follows, with p_1 on the horizontal axis and the expected values on the vertical axis:

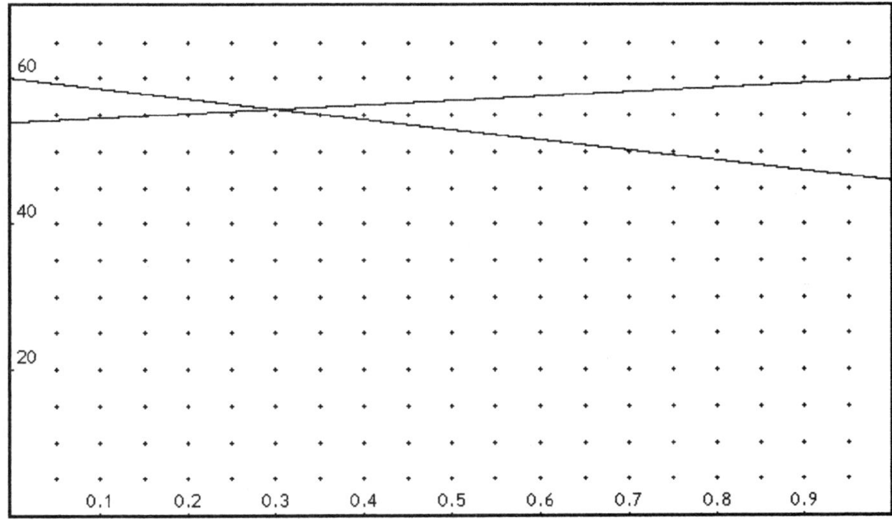

The game's highest expected value is found at the intersection of the two lines above. This is the point where

$$\begin{aligned} 6p_1 + 54 &= -14p_1 + 60 \\ 20p_1 &= 6 \\ p_1 &= \frac{6}{20} \\ &= \frac{3}{10} \end{aligned}$$

The expected value of the game is then $6(\frac{3}{10}) + 54 = 55.8$.

Could 500 Channels! do better with a randomized strategy?

$$\begin{aligned} \begin{bmatrix} p_1 & 1-p_1 \end{bmatrix} \begin{bmatrix} 60 & 46 \\ 54 & 60 \end{bmatrix} \begin{bmatrix} q_1 \\ 1-q_1 \end{bmatrix} &= \begin{bmatrix} 6p_1 + 54 & -14p_1 + 60 \end{bmatrix} \begin{bmatrix} q_1 \\ 1-q_1 \end{bmatrix} \\ &= [(6p_1 + 54)q_1 + (-14p_1 + 60)(1-q_1)] \\ &= [20p_1q_1 - 6q_1 - 14p_1 + 60] \\ &= [20(\tfrac{3}{10})q_1 - 6q_1 - 14(\tfrac{3}{10}) + 60] \\ &= [55.8] \end{aligned}$$

No, it cannot.

If Discs-R-Us follows the strategy found above, it will get 55.8% of the central valley area business regardless of what 500 Channels! does.

3. Here is the reward matrix:

$$R = \begin{bmatrix} 4 & 5 \\ -2 & 0 \end{bmatrix}.$$

Note that row 2 and column 2 are dominated strategies. This leaves row 1 and column 1 as the best pure strategies. (Even without eliminating the dominated strategies we observe that this is a strictly determined game with saddle point 4.) Both stores should have sales; the value of the game is 4% and favors Bloomworthy's.

Chapter 9
Statistics

Section 9.1
Frequency Distributions

1.

No. of children	Frequency	Relative Frequency
0	8	$\frac{8}{40} = \frac{1}{5} = 20\%$
1	13	$\frac{13}{40} = 32.5\%$
2	9	$\frac{9}{40} = 22.5\%$
3	6	$\frac{6}{40} = \frac{3}{20} = 15\%$
4	3	$\frac{3}{40} = 7.5\%$
5	1	$\frac{1}{40} = 2.5\%$

5.

x = Time (min.)	Frequency	Relative Frequency
$0 \le x < 10$	101	$\frac{101}{700} \approx 14.43\%$
$10 \le x < 20$	237	$\frac{237}{700} \approx 33.86\%$
$20 \le x < 30$	169	$\frac{169}{700} \approx 24.14\%$
$30 \le x < 40$	79	$\frac{79}{700} \approx 11.29\%$
$40 \le x < 50$	51	$\frac{51}{700} \approx 7.29\%$
$50 \le x < 60$	63	$\frac{63}{700} = 9\%$

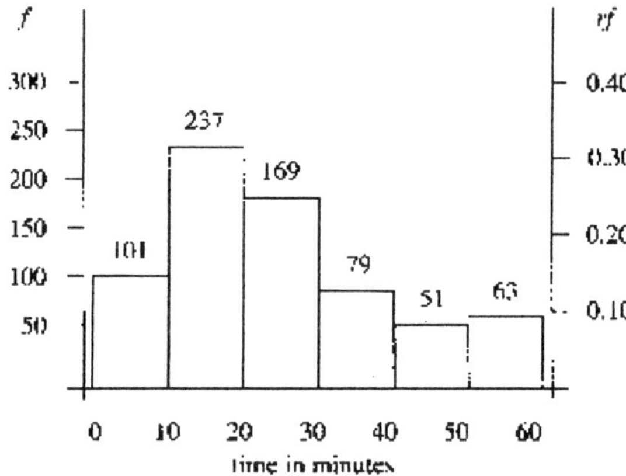

9.

Age	Frequency	Relative Frequency	Group Width	Rel. Freq. Density
$18 \le x < 25$	$13,167,000$	$\frac{13,167,000}{52,585,000} \approx 25.04\%$	7	$\frac{\frac{13,167,000}{52,585,000}}{7} \approx 0.036$
$25 \le x < 30$	$10,839,000$	$\frac{10,839,000}{52,585,000} \approx 20.61\%$	5	$\frac{\frac{10,839,000}{52,585,000}}{5} \approx 0.041$
$30 \le x < 35$	$10,838,000$	$\frac{10,838,000}{52,585,000} \approx 20.61\%$	5	$\frac{\frac{10,838,000}{52,585,000}}{5} \approx 0.041$
$35 \le x < 40$	$9,586,000$	$\frac{9,586,000}{52,585,000} \approx 18.23\%$	5	$\frac{\frac{9,586,000}{52,585,000}}{5} \approx 0.036$
$40 \le x < 45$	$8,155,000$	$\frac{8,155,000}{52,585,000} \approx 15.51\%$	5	$\frac{\frac{8,155,000}{52,585,000}}{5} \approx 0.031$

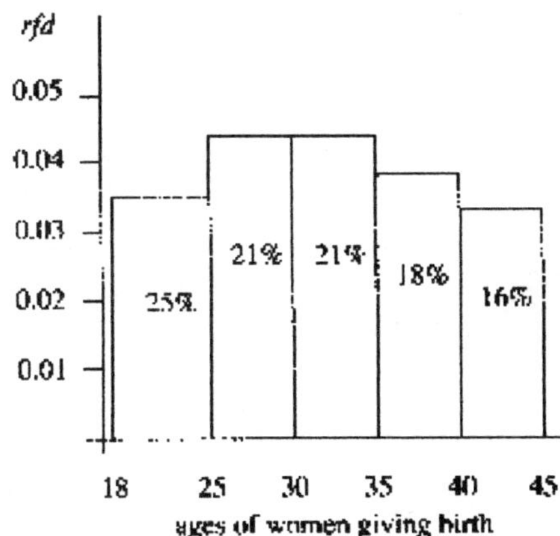

13a. $\frac{52+4+1}{200} = 28.5\%$

13b. "At least 31" means "greater than or equal to 31," or "31 or more." $\frac{16+29+19}{200} = 32\%$

13c. Can't do; can only be approximated by x, where $\frac{1+4}{200} < x < \frac{1+4+52}{200}$; i.e. $2.5\% < x < 28.5\%$.

13d. $\frac{52+48+31+16+29+19}{200} = 97.5\%$

13e. $\frac{52+48}{200} = 50\%$

13f. $\frac{1+4+52+29+19}{200} = 52.5\%$

17a. When the raw data consists of many non-repeated data points.

17b. General trends are easily seen, but some information is lost.

Measures of Central Tendency

1. Note that there are 14 items (data points).
Median $= \frac{\text{Item 7+Item 8}}{2} = \frac{11+12}{2} = \frac{23}{2} = 11.5$. Mode $= 9$.
Mean $= \frac{9+12+8+10+9+11+12+15+20+9+14+15+21+10}{14} = 12.5$.

5a. Mean $= \frac{9+9+10+11+12+15}{6} = 11$. Median $= \frac{10+11}{2} = 10.5$. Mode $= 9$.

5b. Mean $= \frac{9+9+10+11+12+102}{6} = 25.5$. Median $= \frac{10+11}{2} = 10.5$. Mode $= 9$.

5c. Means are different because one of the data points are different.

9. Mean $= 43.933$ (there are 15 data points). To find the median, arrange the data points in order: 22, 25, 34, 35, 41, 41, 46, 46, 46, 47, 49, 54, 54, 59, 60. Median $= 46$ (8th data point). Mode $= 46$.

13.

x = weight (oz.)	f = no. of boxes	y = midpoint	$f \cdot y$
$15.3 \le x < 15.6$	13	$\frac{15.3+15.6}{2} = 15.45$	200.85
$15.6 \le x < 15.9$	24	15.75	378
$15.9 \le x < 16.2$	84	16.05	1348.2
$16.2 \le x < 16.5$	19	16.35	310.65
$16.5 \le x < 16.8$	10	16.65	166.5

$$\begin{aligned} \overline{x} &= \frac{\Sigma(f \cdot y)}{n} \\ &= \frac{200.85 + 378 + 1348.2 + 310.65 + 166.5}{150} \\ &= 16.028 \end{aligned}$$

17. Mean time out $= \frac{90}{60} = 1.5$ hrs. Mean time back $= \frac{90}{45} = 2$ hrs. Round trip total time $= 3.5$ hrs. Mean speed $= \frac{180}{3.5} \approx 51.43$ mph.

21a.

x = Age (yrs.)	f = no. of people	y = midpoint	$f \cdot y$
$0 < x < 5$	18,354,000	2.5	45,885,000
$5 \le x < 18$	42,250,000	11.5	485,875,000
$18 \le x < 21$	11,727,000	19.5	228,676,500
$21 \le x < 25$	15,011,000	23	345,253,000
$25 \le x < 45$	80,755,000	35	2,826,425,000
$45 \le x < 55$	25,223,000	50	1,261,150,000
$55 \le x < 60$	10,532,000	57.5	605,590,000
$60 \le x < 65$	10,616,000	62.5	663,500,000
$65 \le x < 75$	18,107,000	70	1,267,490,000
$75 \le x < 85$	10,055,000	80	804,400,000

$$\begin{aligned}\mu &= \frac{\Sigma(f \cdot y)}{n} \\ &= \frac{8,534,244,500}{242,630,000} \\ &\approx 35.17\end{aligned}$$

21b.

x = Age (yrs.)	f = no. of people	y = midpoint	$f \cdot y$
$0 < x < 5$	18,354,000	2.5	45,885,000
$5 \le x < 18$	42,250,000	11.5	485,875,000
$18 \le x < 21$	11,727,000	19.5	228,676,500
$21 \le x < 25$	15,011,000	23	345,253,000
$25 \le x < 45$	80,755,000	35	2,826,425,000
$45 \le x < 55$	25,223,000	50	1,261,150,000
$55 \le x < 60$	10,532,000	57.5	605,590,000
$60 \le x < 65$	10,616,000	62.5	663,500,000
$65 \le x < 75$	18,107,000	70	1,267,490,000
$75 \le x < 85$	10,055,000	80	804,400,000
$85 \le x \le 100$	3,080,000	92.5	284,900,000

$$\begin{aligned}\mu &= \frac{\Sigma(f \cdot y)}{n} \\ &= \frac{8,534,244,500 + 284,900,000}{242,630,000 + 3,080,000} \\ &\approx 35.89\end{aligned}$$

25. It acts as the representative of the grouped data in that interval. It is the "mean" of the interval.

Section 9.3
Measures of Dispersion

1a. Note that $\overline{x} = 7$.

$$\begin{aligned}
s^2 &= \frac{\Sigma(x-\overline{x})^2}{n-1} \\
&= \frac{(3-7)^2+(8-7)^2+(5-7)^2+(3-7)^2+(10-7)^2+(13-7)^2}{6-1} \\
&= \frac{(-4)^2+(-1)^2+(-2)^2+(-4)^2+3^2+6^2}{5} \\
&= \frac{16+1+4+16+9+36}{5} \\
&= \frac{82}{5} \\
&= 16.4
\end{aligned}$$

$$\begin{aligned}
s &= \sqrt{16.4} \\
&\approx 4.0497
\end{aligned}$$

1b.

$$\begin{aligned}
s^2 &= \frac{1}{n-1}\left[\Sigma x^2 - \frac{(\Sigma x)^2}{n}\right] \\
&= \frac{1}{6-1}\left[\left(3^2+8^2+5^2+3^2+10^2+13^2\right) - \frac{(3+8+5+3+10+13)^2}{6}\right] \\
&= \frac{1}{5}\left[(9+64+25+9+100+169) - \frac{42^2}{6}\right] \\
&= \frac{1}{5}\left(376 - \frac{1764}{6}\right) \\
&= \frac{1}{5}(376-294) \\
&= \frac{1}{5}\cdot 82 \\
&= 16.4
\end{aligned}$$

$$\begin{aligned}
s &= \sqrt{16.4} \\
&\approx 4.0497
\end{aligned}$$

5a.

$$\begin{aligned}
\overline{x} &= 22, \\
s^2 &= \frac{\Sigma(x-\overline{x})^2}{n-1} \\
&= \frac{(12-22)^2+(16-22)^2+(20-22)^2+(24-22)^2+(28-22)^2+(32-22)^2}{6-1} \\
&= \frac{(-10)^2+(-6)^2+(-2)^2+2^2+6^2+10^2}{5} \\
&= \frac{100+36+4+4+36+100}{5} \\
&= \frac{280}{5} \\
&= 56, \\
s &= \sqrt{56} \\
&= 2\sqrt{14} \\
&\approx 7.4833
\end{aligned}$$

5b.

$$\begin{aligned}
\overline{x} &= 1100, \\
s^2 &= \frac{\Sigma(x-\overline{x})^2}{n-1} \\
&= \frac{(600-1100)^2+(800-1100)^2+(1000-1100)^2+(1200-1100)^2+(1400-1100)^2+(1600-1100)^2}{6-1} \\
&= \frac{(-500)^2+(-300)^2+(-100)^2+100^2+300^2+500^2}{5} \\
&= \frac{250,000+90,000+10,000+10,000+90,000+250,000}{5} \\
&= \frac{700,000}{5} \\
&= 140,000, \\
s &= \sqrt{140,000} \\
&= 100\sqrt{14} \\
&\approx 374.1657
\end{aligned}$$

5c. The data in (b) are 50 times the data in (a).

5d. In (b) $\overline{x}$ is 50 times that in (a) and s^2 is 50^2 times that in (a), thus s is 50 times that in (a).

9.

$$
\begin{aligned}
\mu &\approx 3.2083, \\
\sigma &= \sqrt{\frac{\Sigma(x-\mu)^2}{n}} \\
&\approx \sqrt{\frac{(6.0-3.2083)^2+(4.2-3.2083)^2+\cdots+(6.3-3.2083)^2}{12}} \\
&\approx \sqrt{\frac{7.7936+0.9835+0.1534+0.6533+2.5866+3.2699+6.2916+3.6416+1.4600+0.0367+5.7202+9.5586}{12}} \\
&= \sqrt{\frac{42.149}{12}} \\
&\approx \sqrt{3.5124} \\
&\approx 1.8741
\end{aligned}
$$

13.

x = Weight (oz.)	Frequency f	Midpoint y	$f \cdot y$	$f \cdot y^2$
$15.3 \le x < 15.6$	13	15.45	200.85	3103.1325
$15.6 \le x < 15.9$	24	15.75	378	5953.5
$15.9 \le x < 16.2$	84	16.05	1348.2	21,638.61
$16.2 \le x < 16.5$	19	16.35	310.65	5079.1275
$16.5 \le x < 16.8$	10	16.65	166.5	2772.225

$$
\begin{aligned}
n &= 150, \\
\Sigma(f \cdot y) &= 2404.2, \\
\Sigma(f \cdot y^2) &= 38,546.595.
\end{aligned}
$$

$$
\begin{aligned}
s &= \sqrt{\frac{1}{n-1}\left\{\Sigma(f \cdot y^2) - \frac{[\Sigma(f \cdot y)]^2}{n}\right\}} \\
&= \sqrt{\frac{1}{150-1}\left(38,546.595 - \frac{2404.2^2}{150}\right)} \\
&= \sqrt{\frac{1}{149} \cdot 12.0774} \\
&\approx \sqrt{0.0810564} \\
&\approx 0.2847
\end{aligned}
$$

17a. Negative deviations "cancel out" positive deviations.

17b. Decreases it.

17c. Increases it.

17d. It de-emphasizes small deviations (deviations less than 1) and emphasizes large deviations (deviations greater than 1).

17e. To make the units more reasonable.

1a.

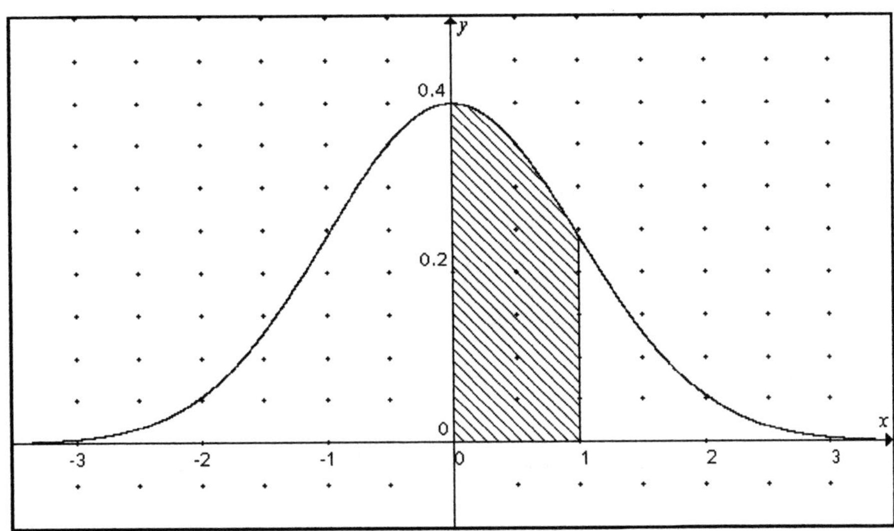

$p(0 < z < 1) \approx 0.3413$

1b.

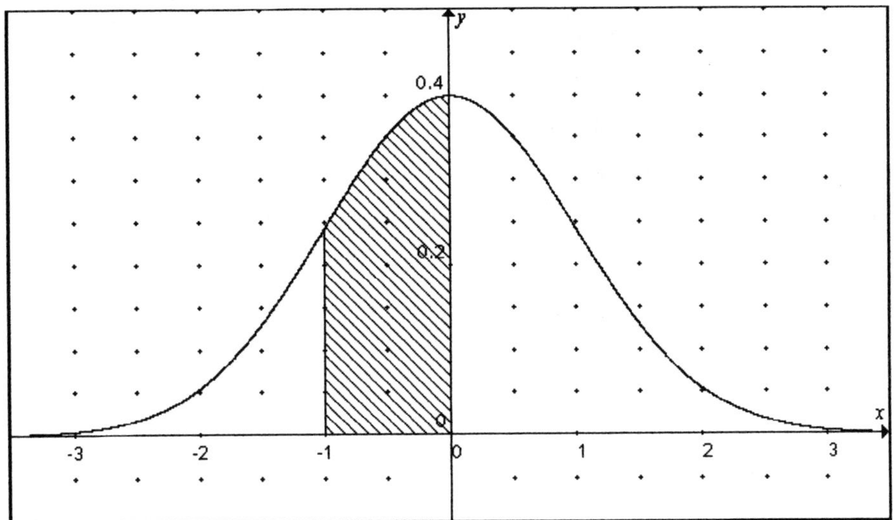

$p(-1 < z < 0) \approx 0.3413$

1c.

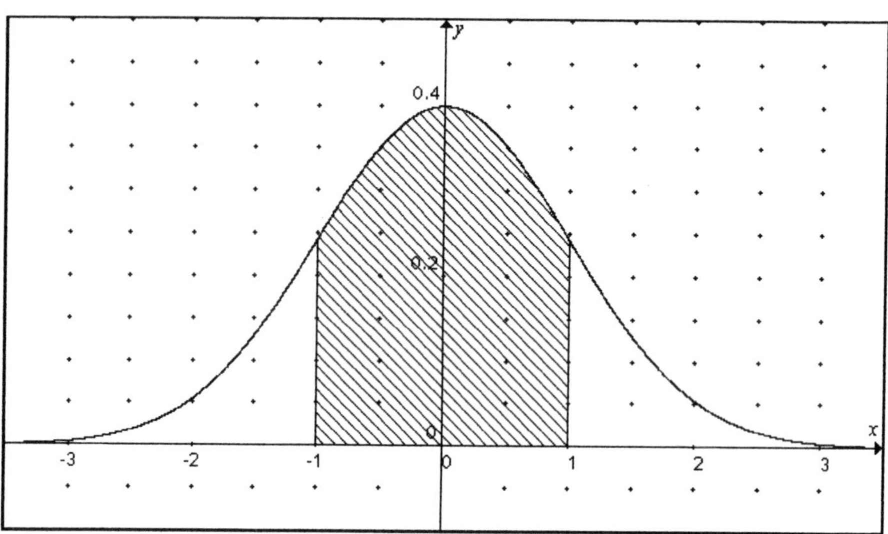

$$p(-1 < z < 1) \approx 2 \cdot 0.3413 = 0.6826$$

5a. One standard deviation of the mean:

$$\begin{aligned}[\mu - \sigma,\ \mu + \sigma] &= [24.7 - 2.3,\ 24.7 + 2.3] \\ &= [22.4,\ 27]\end{aligned}$$

Two standard deviations of the mean:

$$\begin{aligned}[\mu - 2\sigma,\ \mu + 2\sigma] &= [24.7 - 2(2.3),\ 24.7 + 2(2.3)] \\ &= [24.7 - 4.6,\ 24.7 + 4.6] \\ &= [20.1,\ 29.3]\end{aligned}$$

Three standard deviations of the mean:

$$\begin{aligned}[\mu - 3\sigma,\ \mu + 3\sigma] &= [24.7 - 3(2.3),\ 24.7 + 3(2.3)] \\ &= [24.7 - 6.9,\ 24.7 + 6.9] \\ &= [17.8,\ 31.6]\end{aligned}$$

5b. 68.26%, 95.44%, and 99.74%, respectively.

5c.

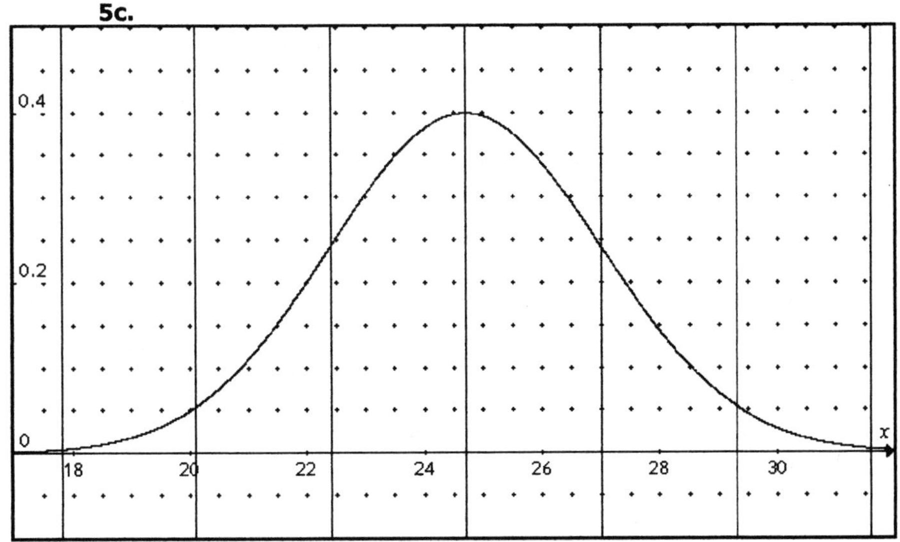

9a. $c = 0.34$

9b. Note that $c < 0$. Since the normal distribution is symmetric, find $p(0 < z < c')$, where $c = -c'$. $c = -2.08$.

9c. Note that this is a strip with width $[-c, c]$; this interval has μ as its midpoint. Thus find c for $p(0 < z < c) = \frac{0.4648}{2} = 0.2324$. $c = 0.62$.

9d. Since $0.6064 > 0.5$, we know that $c < 0$. The narrow strip from c to μ has width $0.6064 - 0.5 = 0.1064$. Looking this number up in the table, we find that $c = -0.27$.

9e. This number (0.0505) is less than 0.5 and c is the lower bound, so c is in the right tail of the distribution. $c = 1.64$.

9f. This number (0.1003) is less than 0.5 and c is the upper bound, so c is in the left tail of the distribution. $c = -1.28$.

13a.

$$\begin{aligned} z &= \frac{x-\mu}{\sigma} \\ &= \frac{15.5-16}{0.3} \\ &\approx -1.67 \end{aligned}$$

$$\begin{aligned} p(x < 15.5) &\approx p(z < -1.67) \\ &= p(z > 1.67) \\ &= 0.0475 \end{aligned}$$

13b.

$$\begin{aligned} p(15.8 < x < 16.2) &= p\left(\frac{15.8-\mu}{\sigma} < z < \frac{16.2-\mu}{\sigma}\right) \\ &= p\left(\frac{15.8-16}{0.3} < z < \frac{16.2-16}{0.3}\right) \\ &\approx p(-0.67 < z < 0.67) \\ &= 2p(0 < z < 0.67) \\ &= 2 \cdot 0.2486 \\ &= 0.4972 \end{aligned}$$

17a.

$$\begin{aligned} p(x \geq 0.9) &= p\left(z \geq \frac{0.9-\mu}{\sigma}\right) \\ &= p\left(z \geq \frac{0.9-1}{0.06}\right) \\ &\approx p(z \geq -1.67) \\ &= 0.5 + p(0 < z < 1.67) \\ &= 0.5 + 0.4525 \\ &= 0.9525 \end{aligned}$$

17b.

$$\begin{aligned} p(x \leq 1.05) &= p\left(z \leq \frac{1.05-\mu}{\sigma}\right) \\ &= p\left(z \leq \frac{1.05-1}{0.06}\right) \\ &\approx p(z \leq 0.833) \\ &= 0.5 + p(0 < z < 0.833) \\ &= 0.5 + 0.2976 \\ &= 0.7976 \end{aligned}$$

21.

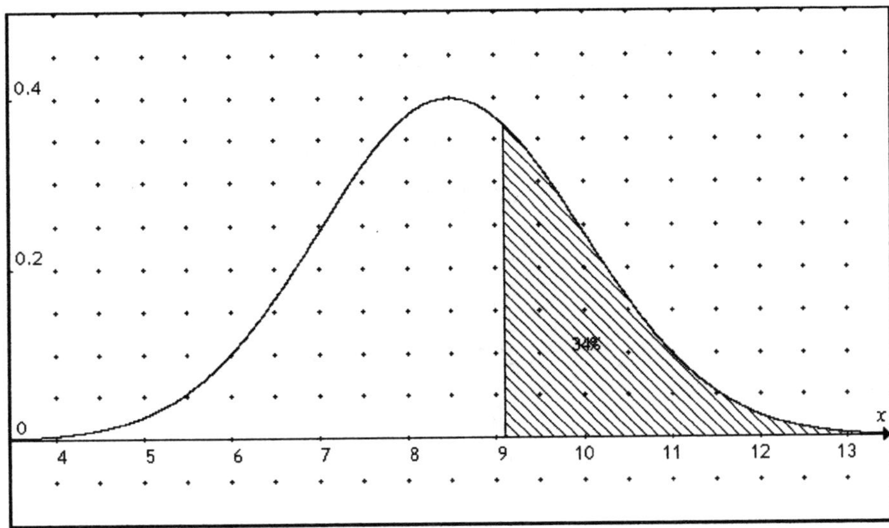

$$
\begin{aligned}
p(x > c) &= 0.34 \\
&= p(z > c') \\
p(0 < z < c') &= 0.5 - 0.34 \\
&= 0.16 \\
\therefore c' &= 0.41 \\
\frac{c - 8.5}{1.5} &= 0.41 \\
c &= 9.115
\end{aligned}
$$

Section 9.5
Binomial Experiments

1. "At least 40 out of 41 times" means 40 or 41 times, thus we seek $p(x=40)+p(x=41)$. We define success to be "Democrat"; $n=41$, $p=\frac{1}{2}$, $q=1-p=\frac{1}{2}$:

$$\begin{aligned} p(x=40)+p(x=41) &= {}_{41}C_{40}p^{40}q^{41-40}+{}_{41}C_{41}p^{41}q^{41-41} \\ &= {}_{41}C_{40}\left(\frac{1}{2}\right)^{40}\left(\frac{1}{2}\right)^{1}+{}_{41}C_{41}\left(\frac{1}{2}\right)^{41}\left(\frac{1}{2}\right)^{0} \\ &= 41\cdot\frac{1}{2^{40}}\cdot\frac{1}{2}+1\cdot\frac{1}{2^{41}}\cdot 1 \\ &= \frac{41}{2^{41}}+\frac{1}{2^{41}} \\ &= \frac{41+1}{2^{41}} \\ &= \frac{42}{2^{41}} \\ &= \frac{2\cdot 21}{2^{41}} \\ &= \frac{21}{2^{40}} \\ &\approx 1.91\times 10^{-11} \end{aligned}$$

5a. Recall that the binomial probability formula is ${}_nC_xp^xq^{n-x}$, where p is the probability of success, q is the probability of failure, and x is the number of successes in n independent trials. In the expression given, $n=18$, so there are 18 trials.

5b. In the expression given, $x=10$, so there are 10 successes.

5c. There are $n-x=18-10=8$ failures.

5d. The probability of success is $p=0.8$.

5e. The probability of failure is $q=0.2$.

9a. Let success be defined as heads; then $p=q=\frac{1}{2}$.

$$\begin{aligned} p(x=0) &= {}_nC_xp^xq^{n-x} \\ &= {}_3C_0\left(\frac{1}{2}\right)^{0}\left(\frac{1}{2}\right)^{3-0} \\ &= 1\cdot 1\cdot\frac{1}{8} \\ &= \frac{1}{8} \end{aligned}$$

9b.

$$\begin{aligned} p(x=1) &= {}_nC_x p^x q^{n-x} \\ &= {}_3C_1 \left(\frac{1}{2}\right)^1 \left(\frac{1}{2}\right)^{3-1} \\ &= 3 \cdot \frac{1}{2} \cdot \frac{1}{4} \\ &= \frac{3}{8} \end{aligned}$$

9c.

$$\begin{aligned} p(x=2) &= {}_nC_x p^x q^{n-x} \\ &= {}_3C_2 \left(\frac{1}{2}\right)^2 \left(\frac{1}{2}\right)^{3-2} \\ &= 3 \cdot \frac{1}{4} \cdot \frac{1}{2} \\ &= \frac{3}{8} \end{aligned}$$

9d.

$$\begin{aligned} p(x=3) &= {}_nC_x p^x q^{n-x} \\ &= {}_3C_3 \left(\frac{1}{2}\right)^3 \left(\frac{1}{2}\right)^{3-3} \\ &= 1 \cdot \frac{1}{8} \cdot 1 \\ &= \frac{1}{8} \end{aligned}$$

9e.

No. of heads	Probability
$x = 0$	$\frac{1}{8}$
$x = 1$	$\frac{3}{8}$
$x = 2$	$\frac{3}{8}$
$x = 3$	$\frac{1}{8}$

9f.

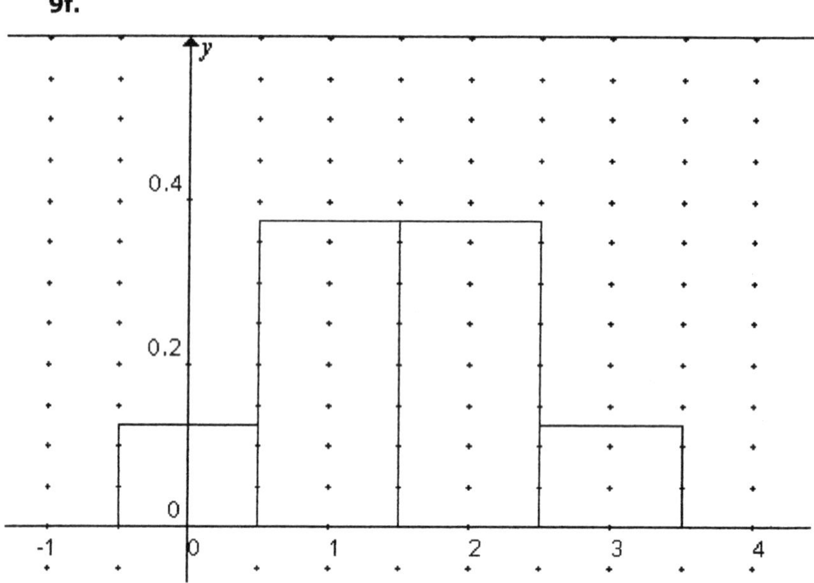

13. The first experiment is not binomial because there are more outcomes than can be classified as "success" or "failure." The second experiment is binomial.

17a. This is a binomial "experiment" because the outcomes can be classified as "success" (the person actually has the disease) or "failure" (the person does not have the disease), the experiment is repeated several times (1000 times in this case), and the trials' outcomes are independent. Recall that the expected number of successes in a binomial experiment with n trials and a probability of success p is np. Here $n = 1000,\ p = 0.000004,$ and $np = 0.004.$

17b.

$$\begin{aligned} p(x=0) &= {}_nC_x p^x q^{n-x} \\ &= {}_{1000}C_0(0.000004)^0(1-0.000004)^{1000-0} \\ &= 1 \cdot 1 \cdot 0.999996^{1000} \\ &\approx 0.996 \end{aligned}$$

17c. $np = 1000 \cdot 0.49 = 490$

21. "If E and F are mutually exclusive, then $p(E \cup F) = p(E) + p(F)$."

25a. (Hint: A probability is a ratio.)

25b. (Hint: Relative frequencies are composed of data from actual experiments.)

29a. $n = 5,\ p = \frac{1}{2},\ q = \frac{1}{2}$

No. of children with Type AB	Probability
0	0.03125
1	0.15625
2	0.3125
3	0.3125
4	0.15625
5	0.03125

29c. 2 or 3

29d. $np = 5 \cdot 0.5 = 2.5$

Section 9.6 The Normal Approximation to the Binomial Distribution

1. We seek $p(x \geq 229.5)$, where $\mu = np = 1000 \cdot 0.2 = 200$ and $\sigma = \sqrt{npq} = \sqrt{(np)q} = \sqrt{\mu(1-p)} = \sqrt{200 \cdot 0.8} = \sqrt{160} = 4\sqrt{10}$.

$$\begin{aligned} p(x > 229.5) &= p\left(z > \frac{229.5 - 200}{4\sqrt{10}}\right) \\ &\approx p(z > 2.33) \\ &= 0.5 - p(0 < z < 2.33) \\ &= 0.5 - 0.4901 \\ &= 0.0099 \end{aligned}$$

Thus in the long run the probability of 230 or more successes is about 1%.

5b. Here the number of "trials" is 1000, and the probability of "success" is 0.51. Thus $\mu = np = 1000(0.51) = 510$ and $\sigma = \sqrt{510(0.49)} = \sqrt{249.9}$.

$$\begin{aligned} p(x > 9.5) &= p\left(z > \frac{9.5 - 510}{\sqrt{249.9}}\right) \\ &\approx p(z > -31.66) \\ &= 0.5 + p(0 < z < 31.66) \\ &> 0.999 \end{aligned}$$

9a. Number of trials is 152 and probability of success is 0.34. We use the Binomial Probability Formula:

$$\begin{aligned} p(x = 0) &= {}_nC_x p^x q^{n-x} \\ &= {}_{152}C_0 (0.34)^0 (0.66)^{152-0} \\ &\approx 1 \cdot 1 \cdot 3.72 \times 10^{-28} \\ &\approx 0 \end{aligned}$$

9b.

$$\begin{aligned} p(x < 45.5) &= p\left(z < \frac{45.5 - 51.68}{\sqrt{34.1088}}\right) \\ &\approx p(z < -1.06) \\ &= p(z > 1.06) \\ &= 0.5 - p(0 < z < 1.06) \\ &= 0.5 - 0.3554 \\ &= 0.1446 \end{aligned}$$

9c.

$$\begin{aligned} p(x > 37.5) &= p\left(z > \frac{37.5 - 51.68}{\sqrt{34.1088}}\right) \\ &\approx p(z > -2.428) \\ &= 0.5 + p(0 < z < 2.428) \\ &\approx 0.9924 \end{aligned}$$

Review Exercises

1a.

No. of Children	Frequency	Relative Frequency
0	8	$\frac{8}{40} = 0.2$
1	10	$\frac{10}{40} = 0.25$
2	11	$\frac{11}{40} = 0.275$
3	7	$\frac{7}{40} = 0.175$
4	3	$\frac{3}{40} = 0.075$
5	1	$\frac{1}{40} = 0.025$

1b. $\overline{x} = \frac{3+1+0+4+1+3+2+2+0+\cdots+2+0+2}{40} = \frac{70}{40} = \frac{7}{4} = 1.75$

1c. Note that when n is even, the median (after the data are sorted into ascending order) is $\frac{a_{n/2}+a_{n/2+1}}{2}$, where the sorted data are denoted $a_1, a_2, \ldots, a_n$. In other words, after the data are sorted in order, a_i is the ith element in the set of data. For example, a_7 is the 7th data item, which, in this exercise, is zero.

In this case we have $n = 40$ and the median is $\frac{a_{n/2}+a_{n/2+1}}{2} = \frac{a_{40/2}+a_{40/2+1}}{2} = \frac{a_{20}+a_{20+1}}{2} = \frac{a_{20}+a_{21}}{2} = \frac{2+2}{2} = 2$, where we used the table with its frequencies from 1a above to get a_{20} and a_{21}.

Also, when n is odd and the sorted data are denoted $a_1, a_2, \ldots, a_n$, the median is the data element $a_{(n+1)/2}$.

More simply, since there are 40 data points, the median is the average of the 20th and 21st data points, when they are arranged in order. Since the 20th data point is 2 children, as is the 21st, the median is $\frac{2+2}{2} = 2$.

1d. From the table in 1a we see that the mode is 2, since this number occurs more often (11 times) than any other number in the data.

1e.

$$\begin{aligned} s &= \sqrt{\frac{\Sigma(x-\overline{x})^2}{n-1}} \\ &= \sqrt{\frac{(3-1.75)^2 + (1-1.75)^2 + (0-1.75)^2 + (4-1.75)^2 + \cdots + (2-1.75)^2}{40-1}} \\ &\approx 1.3156 \end{aligned}$$

2a.

$$\begin{aligned} \frac{23+45}{23+45+53+31+17} &= \frac{68}{169} \\ &\approx 40.24\% \end{aligned}$$

2b.

$$\begin{aligned} \frac{31+17}{23+45+53+31+17} &= \frac{48}{169} \\ &\approx 28.40\% \end{aligned}$$

2c.

$$\frac{45+53+31+17}{23+45+53+31+17} = \frac{146}{169} \approx 86.39\%$$

2d.

$$\frac{23}{23+45+53+31+17} = \frac{23}{169} \approx 13.61\%$$

2e.

$$\frac{53+31}{23+45+53+31+17} = \frac{84}{169} \approx 49.70\%$$

2f. Cannot be determined, but certainly at least $\frac{53+31+17}{23+45+53+31+17} = \frac{101}{169} \approx 59.76\%$ and no more than $\frac{45+53+31+17}{23+45+53+31+17} = \frac{146}{169} \approx 86.39\%$.

3a.

x = Time (min.)	f = Frequency	y = Midpoint	$f \cdot y$
$3 \le x < 6$	18	4.5	81
$6 \le x < 9$	42	7.5	315
$9 \le x < 12$	64	10.5	672
$12 \le x < 15$	35	13.5	472.5
$15 \le x \le 18$	12	16.5	198

$$\begin{aligned}\bar{x} &= \frac{\Sigma(f \cdot y)}{n} \\ &= \frac{81+315+672+472.5+198}{18+42+64+35+12} \\ &= \frac{1738.5}{171} \\ &\approx 10.17 \text{ min.}\end{aligned}$$

3b.

x = Time (min.)	f = Frequency	y = Midpoint	$f \cdot y$	$f \cdot y^2$
$3 \le x < 6$	18	4.5	81	364.5
$6 \le x < 9$	42	7.5	315	2362.5
$9 \le x < 12$	64	10.5	672	7056
$12 \le x < 15$	35	13.5	472.5	6378.75
$15 \le x \le 18$	12	16.5	198	3267

$$\begin{aligned}s &= \sqrt{\frac{1}{n-1}\left[\Sigma(f \cdot y^2) - \frac{\Sigma(f \cdot y)^2}{n}\right]} \\ &= \sqrt{\frac{1}{171-1}\left(19,428.75 - \frac{1738.5^2}{171}\right)} \\ &\approx \sqrt{10.3176470588} \\ &\approx 3.2121\end{aligned}$$

3c.

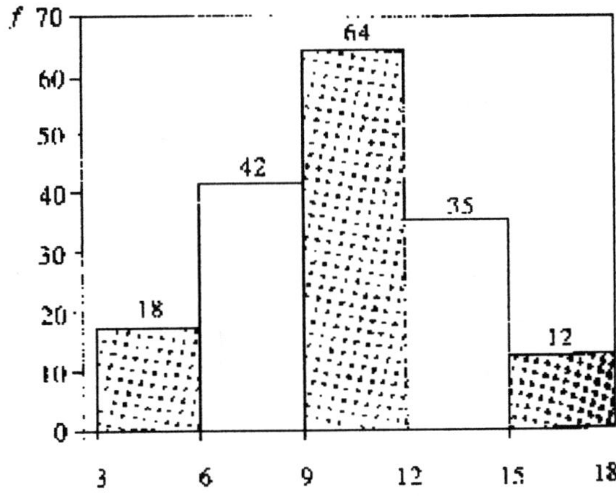

4.

$$\frac{74+65+85+76+x}{5} \geq 80$$
$$\frac{300+x}{5} \geq 80$$
$$300+x \geq 5 \cdot 80$$
$$x \geq 400-300 = 100$$

5. Let x_i, $i = 1, 2, \ldots, 12$, represent the salaries of the 12 men and y_j, $j = 1, 2, \ldots, 8$ represent the salaries of the 8 women. Then

$$\frac{x_1+x_2+\cdots+x_{12}}{12} = 37,000$$
$$x_1+x_2+\cdots+x_{12} = 12 \cdot 37,000$$
$$= 444,000,$$

and

$$\frac{y_1+y_2+\cdots+y_8}{8} = 28,000$$
$$y_1+y_2+\cdots+y_8 = 8 \cdot 28,000$$
$$= 224,000.$$

Thus the mean salary of all 20 people is

$$\frac{(x_1+x_2+\cdots+x_{12})+(y_1+y_2+\cdots+y_8)}{20} = \frac{444,000+224,000}{20}$$
$$= \frac{668,000}{20}$$
$$= \$33,400.$$

6a. Timo: $\frac{103+99+107+93+92}{5} = 98.8$; Henke: $\frac{101+92+83+96+111}{5} = 96.6$

6b. Timo:

$$s = \sqrt{\frac{1}{n-1}\left[\Sigma x^2 - \frac{(\Sigma x)^2}{n}\right]}$$
$$= \sqrt{\frac{1}{4}\left(103^2 + \cdots + 92^2 - \frac{(103 + \cdots + 92)^2}{5}\right)}$$
$$\approx 6.4187$$

Henke:

$$s = \sqrt{\frac{1}{n-1}\left[\Sigma x^2 - \frac{(\Sigma x)^2}{n}\right]}$$
$$= \sqrt{\frac{1}{4}\left(101^2 + \cdots + 111^2 - \frac{(101 + \cdots + 111)^2}{5}\right)}$$
$$\approx 10.4067$$

6c. Timo is the more consistent golfer because his standard deviation is lower (hence his scores cluster more closely about the mean).

7a.

$$\overline{x} = \frac{2(0.99) + 3(1.09) + 7(1.19) + 10(1.29) + 14(1.39) + 4(1.49)}{2 + 3 + 7 + 10 + 14 + 4}$$
$$= \frac{51.9}{40}$$
$$= 1.2975$$

x = **Price/Qt.**	**Frequency** f	$f \cdot x$	$f \cdot x^2$
0.99	2	1.98	1.9602
1.09	3	3.27	3.5643
1.19	7	8.33	9.9127
1.29	10	12.9	16.641
1.39	14	19.46	27.0494
1.49	4	5.96	8.8804

$$s = \sqrt{\frac{1}{n-1}\left[\Sigma(f \cdot x^2) - \frac{\Sigma(f \cdot x)^2}{n}\right]}$$
$$= \sqrt{\frac{1}{40-1}\left(68.008 - \frac{51.9^2}{40}\right)}$$
$$\approx \sqrt{0.0171217948718}$$
$$\approx 0.1309$$

7b. $[\overline{x} - s,\ \overline{x} + s] = [1.2975 - 0.1309,\ 1.2975 + 0.1309] = [1.1666,\ 1.4284]$; $\frac{7+10+14}{40} = 77.5\%$

7c. $[\overline{x} - 2s,\ \overline{x} + 2s] = [1.2975 - 2(0.1309),\ 1.2975 + 2(0.1309)] = [1.0357,\ 1.5593]$; $\frac{3+7+10+14+4}{40} = 95\%$

7d. $[\overline{x} - 3s,\, \overline{x} + 3s] = [1.2975 - 3(0.1309),\ 1.2975 + 3(0.1309)] = [0.9048,$ $1.6902]$; $\frac{2+3+7+10+14+4}{40} = 100\%$

8a. 0.4599, or 45.99%

8b. Same as that between $z = 0$ and $z = 1.75$ because of symmetry: 45.99%

8c. Because of symmetry, twice that between $z = 0$ and $z = 1.75$: 91.98%

9. Let $c =$ the cutoff score; it will be to the left of the mean on the normal curve. It is given that the area of the left tail (or right tail by symmetry) is 0.34; thus the area of the body is 0.5 – 0.34 = 0.16, and this is the number we look up in the table. The closest entry has a z-score of 0.41. Thus

$$\begin{aligned} z &= \frac{x - \mu}{\sigma} \\ 0.41 &= \frac{c - 420}{45} \\ 0.41(45) &= c - 420 \\ 0.41(45) + 420 &= c \\ c &= 438.45 \end{aligned}$$

However, this score is to the right of the mean, i.e., it is an above average score. By symmetry, the score we seek is 420 – (438.45 – 420) = 401.55.

10a. $\frac{1}{2}$; if you picture the normal curve with mean 100, i.e., the curve is split down the middle at 100, then the probability that an adult has a WAIS score $\geq$ 100 is the area of the right half of the normal curve, viz. 0.5.

10b.

$$\begin{aligned} p(x > 125) &= p\left(z > \frac{125 - 100}{15}\right) \\ &\approx p(z > 1.67) \\ &= 0.5 - p(0 < z < 1.67) \\ &= 0.5 - 0.4525 \\ &= 0.0475 \end{aligned}$$

Thus the probability that an adult has a WAIS score of 125 or higher is 4.75%.

10c. Let $c =$ the cutoff score; it will be to the right of the mean on the normal curve. It is given that the area of the right tail is 0.10; thus the area of the body is 0.5 – 0.10 = 0.4, and this is the number we look up in the table. The closest entry has a z-score of 1.28. Thus

$$\begin{aligned} z &= \frac{x - \mu}{\sigma} \\ 1.28 &= \frac{c - 100}{15} \\ 1.28(15) &= c - 100 \\ 1.28(15) + 100 &= c \\ c &= 119.2 \end{aligned}$$

11a. Recall that the expected number of successes in a binomial experiment with n trials and a probability of success p is np. Here $n = 20$ and $p = \frac{1}{5} = 0.2$, where "success" is defined as getting the correct answer. This test taker should expect to get 0.2(20) = 4 correct answers.

11b.

$$\begin{aligned} p(x=20) &= {}_nC_x p^x q^{n-x} \\ &= {}_{20}C_{20}(0.2)^{20}(1-0.2)^{20-20} \\ &\approx 1 \cdot 1.05 \times 10^{-14} \cdot 1 \\ &= 1.05 \times 10^{-14} \end{aligned}$$

11c.

$$\begin{aligned} p(x=18) &= {}_nC_x p^x q^{n-x} \\ &= {}_{20}C_{18}(0.2)^{18}(1-0.2)^{20-18} \\ &\approx (190)\left(2.62 \times 10^{-13}\right)(0.64) \\ &= \left(1.9 \times 10^{2}\right)\left(2.62 \times 10^{-13}\right)\left(6.4 \times 10^{-1}\right) \\ &= (1.9)(2.62)(6.4)\left(10^{2}\right)\left(10^{-13}\right)\left(10^{-1}\right) \\ &= 31.8592 \times 10^{2-13-1} \\ &= 3.18592 \times 10^{1} \times 10^{-12} \\ &\approx 3.19 \times 10^{-11} \end{aligned}$$

12a. Recall that the expected number of successes in a binomial experiment with n trials and a probability of success p is np. Here $n = 20$ and $p = \frac{1}{2} = 0.5$, where "success" is defined as getting the correct answer. This test taker should expect to get 0.5(20) = 10 correct answers.

12b.

$$\begin{aligned} p(x=20) &= {}_nC_x p^x q^{n-x} \\ &= {}_{20}C_{20}(0.5)^{20}(0.5)^{20-20} \\ &\approx 1 \cdot 9.54 \times 10^{-7} \cdot 1 \\ &= 9.54 \times 10^{-7} \end{aligned}$$

12c.

$$\begin{aligned} p(x=18) &= {}_nC_x p^x q^{n-x} \\ &= {}_{20}C_{18}(0.5)^{18}(0.5)^{20-18} \\ &\approx \left(1.9 \times 10^{2}\right)\left(3.81 \times 10^{-6}\right)\left(2.5 \times 10^{-1}\right) \\ &= (1.9)(3.81)(2.5)\left(10^{2}\right)\left(10^{-6}\right)\left(10^{-1}\right) \\ &= 18.0975 \times 10^{2-6-1} \\ &\approx 1.8 \times 10^{1} \times 10^{-5} \\ &= 1.8 \times 10^{-4} \end{aligned}$$

Chapter 10
Finance

Section 10.1
Simple Interest

1.

$$\begin{aligned} I &= Prt \\ &= (2000)(0.08)(3) \\ &= 480 \end{aligned}$$

5.

$$\begin{aligned} I &= Prt \\ &= (1410)(0.1225)\left(\frac{325}{365}\right) \\ &= 153.80 \end{aligned}$$

9.

$$\begin{aligned} FV &= P(1+rt) \\ &= 12,430\left[1+0.05875\left(2+\frac{3}{12}\right)\right] \\ &= 12,430(1.1321875) \\ &= 14,073.09 \end{aligned}$$

13.

$$\begin{aligned} FV &= P(1+rt) \\ &= 5900\left(1+0.145\cdot\frac{112}{365}\right) \\ &\approx 5900(1.04449315068) \\ &= 6162.51 \end{aligned}$$

17.

$$\begin{aligned} FV &= P(1+rt) \\ P &= \frac{FV}{1+rt} \\ &= \frac{8600}{1+0.095(3)} \\ &= 6692.61 \end{aligned}$$

21.

$$\begin{aligned} FV &= P(1+rt) \\ P &= \frac{FV}{1+rt} \\ &= \frac{1311}{1+0.065\cdot\frac{317}{365}} \\ &= \frac{1311}{1+\frac{0.065(317)}{365}} \\ &= \frac{1311}{\frac{365}{365}+\frac{20.605}{365}} \\ &= \frac{1311}{\frac{365+20.605}{365}} \\ &= \frac{1311}{\frac{385.605}{365}} \\ &= 1311\cdot\frac{365}{385.605} \\ &= 1240.95 \end{aligned}$$

25.

$$\begin{aligned} FV &= P(1+rt) \\ &= P+Prt \\ FV-P &= Prt \\ t &= \frac{FV-P}{Pr} \\ &= \frac{1615-1312.82}{1312.82(0.06875)} \\ &\approx 3.35 \text{ yrs.} \end{aligned}$$

29.

$$\begin{aligned} \frac{FV}{36} &= \frac{P(1+rt)}{36} \\ &= \frac{(3700-500)\left(1+0.098\cdot\frac{36}{12}\right)}{36} \\ &= \frac{3200(1.294)}{36} \\ &= 115.02 \end{aligned}$$

33a.

$$\begin{aligned} D &= FV\cdot r\cdot t \\ &= 1750(0.11875)(1.5) \\ &= 311.72 \end{aligned}$$

33b. $1750 - 311.72 = 1438.28$

33c.

$$\begin{aligned} I &= P \cdot r \cdot t \\ r &= \frac{I}{P \cdot t} \\ &= \frac{311.72}{(1438.28)(1.5)} \\ &= 14.45\% \end{aligned}$$

37a. $0.05(162{,}500) = 8125$

37b. $0.90(162{,}500) = 146{,}250$

37c. $8125, which was the amount of the promissory note (5%) due in 4 years.

37d. The future value of the loan is

$$\begin{aligned} FV &= P(1 + rt) \\ &= 8125(1 + 0.10 \cdot 4) \\ &= 8125(1.4) \\ &= 11,375 \end{aligned}$$

Since the principal is 8125, the interest which the Hamiltons will pay is $11{,}375 - 8125 = 3250$. Thus the monthly interest payment (over 4 years = 48 months) is $\frac{3250}{48} = \$67.71$.

37e.
$8125 + (8125 + 48 \cdot 67.71) = 8125 + (8125 + 3250.08) = 8125 + 11{,}375.08$
$= 19{,}500.08$

37f. The bank paid Gurney 90% of the purchase price, or $146,250, less 6% of the purchase price as a sales commission, or $0.06(162{,}500) = 9750$. Thus Gurney received $146{,}250 - 9750 = 136{,}500$ from the Hamiltons' bank.

37g. Gurney's total income from all aspects of the sale consists of the total income from the down payment and income from the Hamiltons' bank: $19{,}500.08 + 136{,}500 = 156{,}000.08$.

41. Let D be the discount (in the Simple Discount Formula). Then

$$\begin{aligned} P &= M - D \\ &= FV - D \\ &= FV - FV \cdot r \cdot t \\ &= FV(1 - r \cdot t) \\ &= M(1 - r \cdot t) \end{aligned}$$

45.

$$\begin{aligned} P &= FV(1 - r \cdot t) \\ &= 3210\left[1 - 0.12375\left(1 + \frac{11}{12}\right)\right] \\ &= 3210\left(1 - 0.12375 \cdot \frac{23}{12}\right) \\ &= 3210(0.7628125) \\ &= \$2448.63 \end{aligned}$$

1a. $\frac{0.12}{4} = 0.03$

1b. $\frac{0.12}{12} = 0.01$

1c. $\frac{0.12}{365} \approx 0.000328767$

1d. $\frac{0.12}{\frac{52}{2}} = 0.12 \cdot \frac{2}{52} \approx 0.004615385$

1e. $\frac{0.12}{2(12)} = \frac{0.12}{24} = 0.005$

5a. $\frac{0.097}{4} = 0.02425$

5b. $\frac{0.097}{12} \approx 0.008083333$

5c. $\frac{0.097}{365} \approx 0.000265753$

5d. $\frac{0.097}{\frac{52}{2}} = 0.097 \cdot \frac{2}{52} \approx 0.003730769$

5e. $\frac{0.097}{2(12)} = \frac{0.097}{24} \approx 0.004041667$

9a. $30\,yr \cdot \frac{4\,qtr}{1\,yr} = 120\,qtr$

9b. $30\,yr \cdot \frac{12\,mo}{1\,yr} = 360\,mo$

9c. $30\,yr \cdot \frac{365\,d}{1\,yr} = 10,950\,d$

13. The future value FV of a principal P placed at an interest i for n periods is $FV = P(1+i)^n$.

$$\begin{aligned} FV &= P(1+i)^n \\ &= 5200\left(1+\frac{0.0675}{4}\right)^{(8.5)(4)} \\ &= 5200(1.016875)^{34} \\ &\approx 5200(1.766433895) \\ &= 9185.46 \end{aligned}$$

17. The nominal rate, or compound rate, is 8%, $n = 1$, and $q = 12$:

$$\begin{aligned} P\left(1+\frac{0.08}{12}\right)^{12} &= P(1+r\cdot 1) \\ \frac{1}{P}\cdot P\left(1+\frac{0.08}{12}\right)^{12} &= \frac{1}{P}\cdot P(1+r) \\ \left(1+\frac{0.08}{12}\right)^{12} &= 1+r \\ r &= \left(1+\frac{0.08}{12}\right)^{12} - 1 \\ &\approx 8.30\% \end{aligned}$$

21a. The nominal rate, or compound rate, is 10%, $n = 1$, and $q = 4$:

$$\begin{aligned}
P\left(1+\frac{0.10}{4}\right)^4 &= P(1+r\cdot 1)\\
\frac{1}{P}\cdot P\left(1+\frac{0.10}{4}\right)^4 &= \frac{1}{P}\cdot P(1+r)\\
\left(1+\frac{0.10}{4}\right)^4 &= 1+r\\
r &= \left(1+\frac{0.10}{4}\right)^4 - 1\\
&\approx 10.38\%
\end{aligned}$$

21b.

$$\begin{aligned}
r &= \left(1+\frac{0.10}{12}\right)^{12} - 1\\
&\approx 10.47\%
\end{aligned}$$

21c.

$$\begin{aligned}
r &= \left(1+\frac{0.10}{365}\right)^{365} - 1\\
&\approx 10.52\%
\end{aligned}$$

25.

$$\begin{aligned}
FV &= P(1+i)^n\\
P &= \frac{FV}{(1+i)^n}\\
&= \frac{3758}{\left(1+\frac{0.11875}{12}\right)^{(17+\frac{7}{12})12}}\\
&\approx \frac{3758}{(1.00989583333)^{(17\cdot 12+\frac{7}{12}\cdot 12)}}\\
&= \frac{3758}{(1.00989583333)^{(204+7)}}\\
&= \frac{3758}{(1.00989583333)^{(211)}}\\
&\approx \frac{3758}{7.98653609626}\\
&\approx 470.54
\end{aligned}$$

29.

$$\begin{aligned}
r_{4\,year} &= \left(1+\frac{0.06}{365}\right)^{365} - 1\\
&\approx 1.00016438356^{365} - 1\\
&\approx 0.06183131004
\end{aligned}$$

$$\begin{aligned}
r_{5\,year} &= \left(1+\frac{0.065}{365}\right)^{365} - 1\\
&\approx 1.00017808219^{365} - 1\\
&\approx 0.06715284808
\end{aligned}$$

33a.

$$\begin{aligned} FV &= P(1+i)^n \\ &= 3000\left(1+\frac{0.065}{365}\right)^{(18)(365)} \\ &= 9664.97 \end{aligned}$$

33b. When he becomes 18 the account balance is \$9664.97, and every 30 days thereafter the interest continues to accrue, but he will withdraw just the interest at the end of every 30-day month, leaving \$9664.97 in the account. Thus we need to find the interest which accrues every thirty days. That is, we find the future value of \$9664.97 after 30 days, and subtract \$9664.97 from it.

$$\begin{aligned} FV - 9664.97 &= P(1+i)^n - 9664.97 \\ &= 9664.97\left(1+\frac{0.065}{365}\right)^{30} - 9664.97 \\ &\approx 9716.74 - 9664.97 \\ &= 51.77 \end{aligned}$$

37a.

$$\begin{aligned} P &= \frac{FV}{(1+i)^n} \\ &= \frac{100,000}{\left(1+\frac{0.08375}{365}\right)^{(65-35)365}} \\ &= 8108.87 \end{aligned}$$

37b. When she retires the account balance is \$100,000, and every 30 days thereafter the interest continues to accrue, but she will withdraw just the interest at the end of every 30-day month, leaving \$100,000 in the account. Thus we need to find the interest which accrues every thirty days. That is, we find the future value of \$100,000 after 30 days, and subtract \$100,000 from it.

$$\begin{aligned} FV - 100,000 &= P(1+i)^n - 100,000 \\ &= 100,000\left(1+\frac{0.08375}{365}\right)^{30} - 100,000 \\ &\approx 100,690.65 - 100,000 \\ &= 690.65 \end{aligned}$$

41. Note that $t = 1$.

$$\begin{aligned} P(1+i)^n &= P(1+rt) \\ \frac{1}{p}\cdot P(1+i)^n &= \frac{1}{P}\cdot P(1+r\cdot 1) \\ (1+i)^n &= 1+r \\ r &= (1+i)^n - 1 \end{aligned}$$

45a.

$$\begin{aligned} r &= (1+i)^n - 1 \\ &= (1+0.15625)^2 - 1 \\ &= 16.24\% \end{aligned}$$

45b.

$$\begin{aligned} r &= (1+i)^n - 1 \\ &= (1+0.15625)^4 - 1 \\ &= 16.56\% \end{aligned}$$

45c.

$$\begin{aligned} r &= (1+i)^n - 1 \\ &= (1+0.15625)^{12} - 1 \\ &= 16.79\% \end{aligned}$$

45d.

$$\begin{aligned} r &= (1+i)^n - 1 \\ &= (1+0.15625)^{365} - 1 \\ &= 16.91\% \end{aligned}$$

45e.

$$\begin{aligned} r &= (1+i)^n - 1 \\ &= (1+0.15625)^{26} - 1 \\ &= 16.86\% \end{aligned}$$

45f.

$$\begin{aligned} r &= (1+i)^n - 1 \\ &= (1+0.15625)^{24} - 1 \\ &= 16.85\% \end{aligned}$$

57a. We need to graph the equations $y = 2$ and $y = \left(1 + \frac{0.05}{1}\right)^x = 1.05^x$, where y measures money and x measures time in years. First we find the proper size for the viewing window. Neither x nor y can be negative, so $x, y \geq 0$. From the text, we recall that the doubling time for 5% simple interest is 20 years, and since simple interest is not as productive as compound interest, the doubling time for 5% compound interest will be less than 20 years. Therefore $x \leq 20$ and the maximum value of y will not be greater than $1.05^{20} \approx 2.65$. The graph appears as follows:

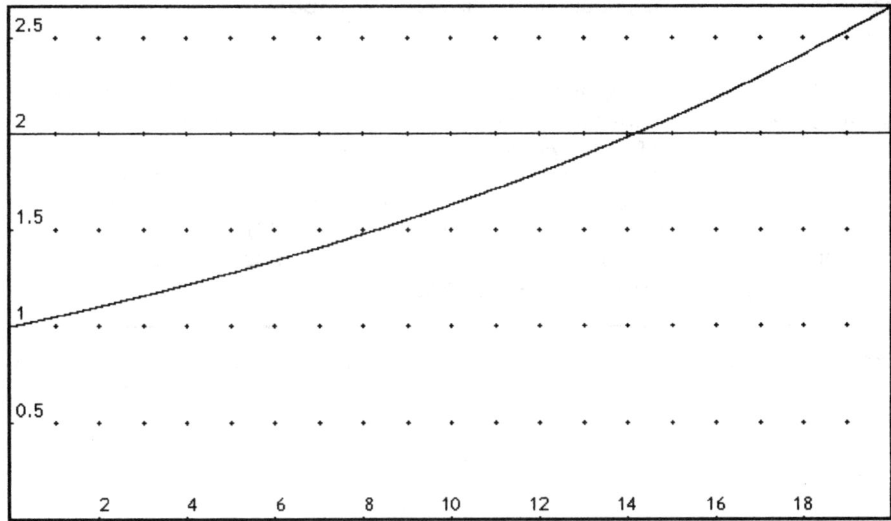

Using the procedure for finding the point of intersection of two graphs on your graphing calculator, you should arrive at (14.2066990829, 2), or a doubling time of about 14.21 years.

57b. We need to graph the equations $y = 2$ and $y = \left(1 + \frac{0.05}{12}\right)^x$, where y measures money and x measures time in months. First we find the proper size for the viewing window. Neither x nor y can be negative, so $x, y \geq 0$. From the text, we recall that the doubling time for 5% simple interest is 20 years, and since simple interest is not as productive as compound interest, the doubling time for 5% compound interest will be less than 20 years. Therefore $x \leq 20 \cdot 12 = 240$ and the maximum value of y will not be greater than $\left(1 + \frac{0.05}{12}\right)^{240} \approx 2.71$. The graph appears as follows:

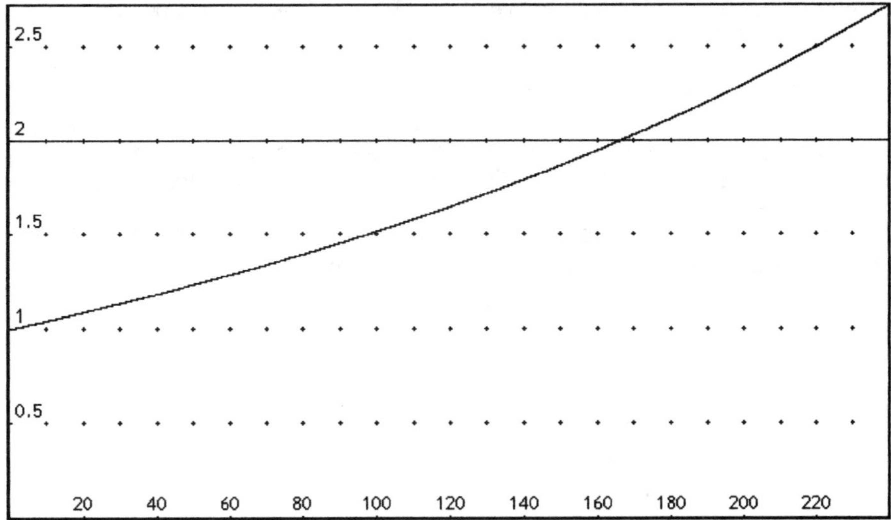

Using the procedure for finding the point of intersection of two graphs on your graphing calculator, you should arrive at (166.701656615, 2), or a doubling time of $\frac{166.701656615}{12} \approx 13.89$ years.

57c. We need to graph the equations $y = 2$ and $y = \left(1 + \frac{0.05}{4}\right)^x$, where y measures money and x measures time in quarters (i.e., 3-month periods). First we find the proper size for the viewing window. Neither x nor y can be negative, so $x, y \geq 0$. From the text, we recall that the doubling time for 5% simple interest is 20 years, and since simple interest is not as productive as compound interest, the doubling time for 5% compound interest will be less than 20 years. Therefore $x \leq 20 \cdot 4 = 80$ and the maximum value of y will not be greater than $\left(1 + \frac{0.05}{4}\right)^{80} \approx 2.70$. The graph appears as follows:

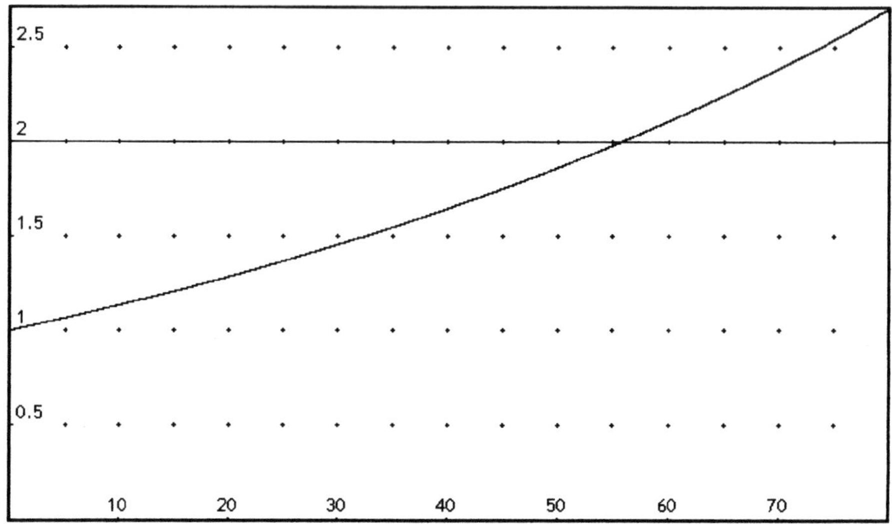

Using the procedure for finding the point of intersection of two graphs on your graphing calculator, you should arrive at (55.797630484, 2), or a doubling time of $\frac{55.797630484}{4} \approx 13.95$ years.

57d. We need to graph the equations $y = 2$ and $y = \left(1 + \frac{0.05}{365}\right)^x$, where y measures money and x measures time in days. First we find the proper size for the viewing window. Neither x nor y can be negative, so $x, y \geq 0$. From the text, we recall that the doubling time for 5% simple interest is 20 years, and since simple interest is not as productive as compound interest, the doubling time for 5% compound interest will be less than 20 years. Therefore $x \leq 20 \cdot 365 = 7300$ and the maximum value of y will not be greater than $\left(1 + \frac{0.05}{365}\right)^{7300} \approx 2.72$. The graph appears as follows:

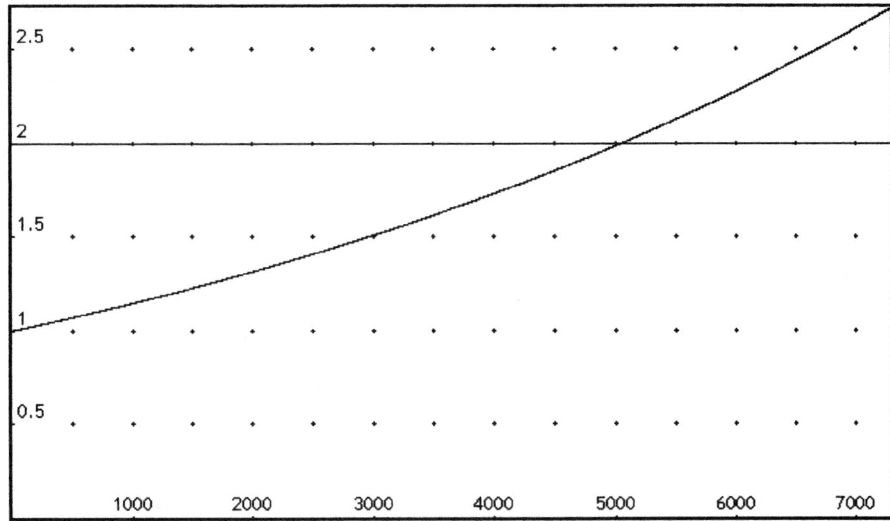

Using the procedure for finding the point of intersection of two graphs on your graphing calculator, you should arrive at (5060.32103435, 2), or a doubling time of $\frac{5060.32103435}{365} \approx 13.86$ years.

61. We need to graph the equations $y = 30,000$, $y = 100,000$, and $y = 20,000(1 + \frac{0.0625}{365})^x$, where y measures money and x measures time in days. First we find the proper size for the viewing window. Neither x nor y can be negative, so $x, y \geq 0$. From the Simple Interest Future Value Formula we obtain that

$$\begin{aligned} FV &= P(1 + rt) \\ 100,000 &= 20,000(1 + 0.0625t) \\ \frac{100,000}{20,000} &= 1 + 0.0625t \\ 5 - 1 &= 0.0625t \\ t &= \frac{4}{0.0625} = 64 \end{aligned}$$

The time for a future value of \$100,000 is 64 years. Therefore $x \leq 64 \cdot 365 = 23,360$ and the maximum value of y will not be greater than $20,000(1 + \frac{0.0625}{365})^{23,360} \approx 1,091,589.2$. The graph appears as follows:

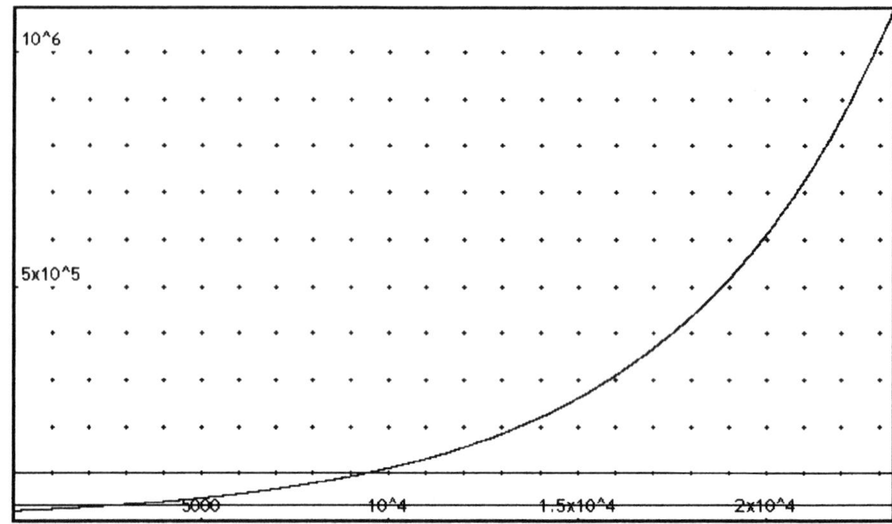

A better view can be obtained by using the ZOOM feature of your graphing calculator. Here is a view with viewing window [xmin, xmax, ymin, ymax] $= [268.07,\ 11,000,\ 0,\ 110,000]$:

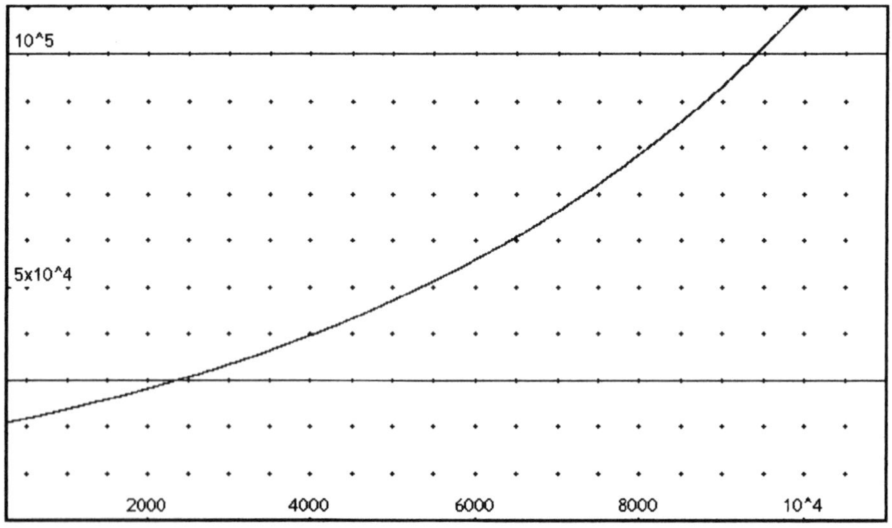

Using the procedure for finding the point of intersection of two graphs on your graphing calculator, you should arrive at (2368.11891264, 30,000), or a time of $\frac{2368.11891264}{365} \approx 6.49$ years, and (9399.92192415, 100,000), or a time of $\frac{9399.92192415}{365} \approx 25.75$ years.

1.

$$\begin{aligned} FV(ordinary) &= pymt\frac{(1+i)^n - 1}{i} \\ &= 120\frac{\left(1+\frac{0.0575}{12}\right)^{12} - 1}{\frac{0.0575}{12}} \\ &\approx 120\frac{0.05903983132}{0.004791667} \\ &= \$1478.56 \end{aligned}$$

5a. This is an ordinary annuity because each payment is due at the end of its time period. The first payment is at the end of February and the last payment is at the end of November, so $n = 10$. Also $i = \frac{0.07}{12}$.

$$\begin{aligned} FV(ordinary) &= pymt\frac{(1+i)^n - 1}{i} \\ &= 75\frac{\left(1+\frac{0.07}{12}\right)^{10} - 1}{\frac{0.07}{12}} \\ &\approx 75\frac{0.05988864761}{0.0058333} \\ &= \$770.00 \end{aligned}$$

5b. He made a total of ten payments of \$75 each, so his total contribution is 10(\$75) = \$750.

5c. Total interest is \$770 – \$750 = \$20.

9a. This is an ordinary annuity with $i = \frac{0.105}{12}$ and $n = (65 - 39)yr. \cdot \frac{12\,mo.}{1\,yr.} = 312$ monthly payments.

$$\begin{aligned} FV(ordinary) &= pymt\frac{(1+i)^n - 1}{i} \\ &= 175\frac{\left(1+\frac{0.105}{12}\right)^{312} - 1}{\frac{0.105}{12}} \\ &\approx 175\frac{14.1518928968}{0.00875} \\ &= \$283,037.86 \end{aligned}$$

9b. $312(175) = 54,600$

9c. $283,037.86 - 54,600 = 228,437.86$

13. We have already found that the future value of the annuity is $\$1478.56$, $i = \frac{0.0575}{12}$, and $n = 12$.

$$\begin{aligned} FV &= P(1+i)^n \\ P &= \frac{FV}{(1+i)^n} \\ &= \frac{1478.56}{\left(1+\frac{0.0575}{12}\right)^{12}} \\ &\approx \$1396.13 \end{aligned}$$

17. We have already found that the future value of the annuity is $\$283,037.86$, $i = \frac{0.105}{12}$, and $n = 312$.

$$\begin{aligned}
FV &= P(1+i)^n \\
P &= \frac{FV}{(1+i)^n} \\
&= \frac{283,037.86}{\left(1+\frac{0.105}{12}\right)^{312}} \\
&\approx \$18,680.03
\end{aligned}$$

21. Note that $i = \frac{0.105}{12}$, $n = 40\,yr. \cdot \frac{12\,mo.}{1\,yr.} = 480$.

$$\begin{aligned}
FV(ordinary) &= pymt\frac{(1+i)^n-1}{i} \\
FV\left[\frac{(1+i)^n-1}{i}\right]^{-1} &= pymt\left[\frac{(1+i)^n-1}{i}\right]^1\left[\frac{(1+i)^n-1}{i}\right]^{-1} \\
FV\frac{i}{(1+i)^n-1} &= pymt\left[\frac{(1+i)^n-1}{i}\right]^{1-1} \\
&= pymt\left[\frac{(1+i)^n-1}{i}\right]^{0} \\
&= pymt\cdot 1
\end{aligned}$$

$$\begin{aligned}
\therefore pymt &= FV\frac{i}{(1+i)^n-1} \\
&= 250,000\frac{\frac{0.105}{12}}{\left(1+\frac{0.105}{12}\right)^{480}-1} \\
&\approx 250,000\frac{0.00875}{64.4791315096} \\
&= 33.93
\end{aligned}$$

25a.

$$\begin{aligned}
FV(ordinary) &= pymt\frac{(1+i)^n-1}{i} \\
&= 100\frac{\left(1+\frac{0.08125}{26}\right)^{923}-1}{\frac{0.08125}{26}} \\
&\approx 100\frac{16.8120916233}{0.003125} \\
&= \$537,986.93
\end{aligned}$$

25b. The Interest column is computed by using the formula for compound interest, $FV = P(1+i)^n$, where $P =$ the account balance at the beginning of the month, $i = \frac{0.061}{12}$, and $n = 1$, and subtracting P from it; i.e.,

$P(1+i)^n - P = P(1+\frac{0.061}{12})^1 - P = P\left[(1+\frac{0.061}{12}) - 1\right] = \frac{0.061}{12}P.$

Mo. No.	Beginning Acct. Balance	Interest	Withdrawal	End Acct. Balance
1	537, 986.93	2734.77	650	540, 071.70
2	540, 071.70	2745.36	650	542, 167.06
3	542, 167.06	2756.02	650	544, 273.08
4	544, 273.08	2766.72	650	546, 389.80
5	546, 389.80	2777.48	650	548, 517.28

29. (See Exercise 21 above for the derivation of the following equation.)

$$\begin{aligned} pymt &= FV\frac{i}{(1+i)^n - 1} \\ &= 1200\frac{\frac{0.09}{12}}{\left(1+\frac{0.09}{12}\right)^{24} - 1} \\ &\approx 1200\frac{0.0075}{0.19641352939} \\ &= 45.82 \end{aligned}$$

33.

$$\begin{aligned} P(1+i)^n &= pymt\frac{(1+i)^n - 1}{i} \\ P(1+i)^n \cdot \frac{1}{(1+i)^n} &= pymt\frac{(1+i)^n - 1}{i} \cdot \frac{1}{(1+i)^n} \\ P &= pymt\frac{(1+i)^n - 1}{i(1+i)^n} \end{aligned}$$

37.

$$\begin{aligned} P &= pymt\frac{(1+i)^n - 1}{i(1+i)^n} \\ &= 75\frac{\left(1+\frac{0.07}{12}\right)^{10} - 1}{\frac{0.07}{12}\left(1+\frac{0.07}{12}\right)^{10}} \\ &\approx 75\frac{0.05988864761}{0.00618268377772} \\ &= 726.49 \end{aligned}$$

41. Since this is an ordinary annuity, we use the formula $FV(ordinary) = pymt\frac{(1+i)^x - 1}{i}$, where x measures time in months (normally this variable would be n), $pymt = 200$, and $i = \frac{0.06}{12}$ (and subsequently $\frac{0.08}{12}$, $\frac{0.10}{12}$). That is, we graph $y_1 = 200\frac{(1+\frac{0.06}{12})^x - 1}{\frac{0.06}{12}} = 200 \cdot \frac{12}{0.06}\left[(1+\frac{0.06}{12})^x - 1\right] = 40,000\left[(1+\frac{0.06}{12})^x - 1\right]$,

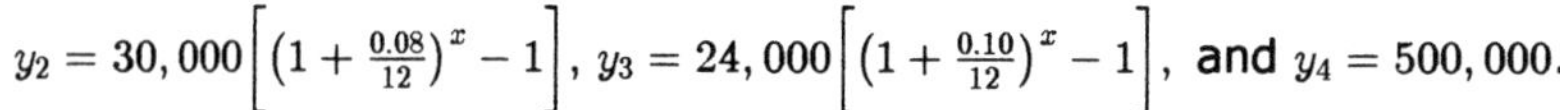

$y_2 = 30,000\left[\left(1 + \frac{0.08}{12}\right)^x - 1\right]$, $y_3 = 24,000\left[\left(1 + \frac{0.10}{12}\right)^x - 1\right]$, and $y_4 = 500,000$.

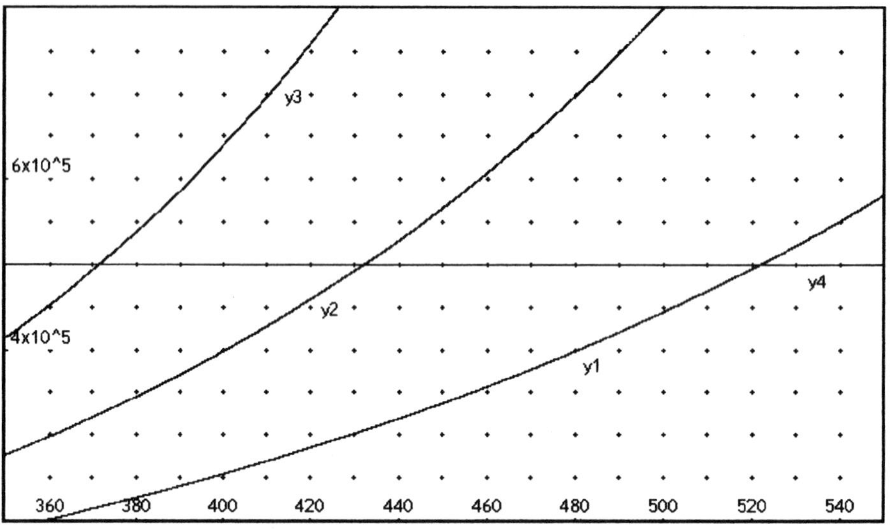

1a.

$$\begin{aligned} pymt\frac{(1+i)^n-1}{i} &= P(1+i)^n \\ pymt\frac{\left(1+\frac{0.095}{12}\right)^{4\cdot 12}-1}{\frac{0.095}{12}} &= 5000\left(1+\frac{0.095}{12}\right)^{4\cdot 12} \\ pymt(58.1176731928) &= 7300.49123055 \\ pymt &= \frac{7300.49123055}{58.1176731928} \\ &= 125.62 \end{aligned}$$

1b. $48 \cdot 125.62 - 5000 = 1029.76$

5a.

$$\begin{aligned} pymt\frac{(1+i)^n-1}{i} &= P(1+i)^n \\ pymt\frac{\left(1+\frac{0.095}{12}\right)^{30\cdot 12}-1}{\frac{0.095}{12}} &= 155,000\left(1+\frac{0.095}{12}\right)^{30\cdot 12} \\ pymt(2033.03517691) &= 2,649,703.58167 \\ pymt &= \frac{2,649,703.58167}{2033.03517691} \\ &= 1303.32 \end{aligned}$$

5b. $(30 \cdot 12)1303.32 - 155,000 = 314,195.20$

9a. Note that $P = 212{,}500 - 0.2(212{,}500) = 170{,}000$.

$$\begin{aligned} pymt\frac{(1+i)^n-1}{i} &= P(1+i)^n \\ pymt\frac{\left(1+\frac{0.10875}{12}\right)^{30\cdot 12}-1}{\frac{0.10875}{12}} &= 170,000\left(1+\frac{0.10875}{12}\right)^{30\cdot 12} \\ pymt(2729.24748279) &= 4,374,746.90318 \\ pymt &= \frac{4,374,746.90318}{2729.24748279} \\ &= 1602.91 \end{aligned}$$

9b. $(30 \cdot 12)1602.91 - 170,000 = 407,047.60$

9c.

Payment No.	Principal Por.	Interest Por.	Tot. Payment	Balance
0				170,000.00
1	$1602.91 - 1540.63 = 62.28$	$170,000(0.10875)(\frac{1}{12}) = 1540.63$	1602.91	169,937.72
2	62.85	1540.06	1602.91	169,874.87

9d. Assuming that his only monthly payments are his house payments, the total of his monthly payments is 1602.91. Let $x =$ his monthly income. Then

$$\begin{aligned} 0.38x &\ge 1602.91 \\ x &\ge \frac{1602.91}{0.38} \\ x &\ge 4218.18. \end{aligned}$$

Thus he must make at least $4218.18 per month.

13a. The car dealer's loan is a 4-year add-on interest loan at 7.75% requiring a down payment of $1000. Thus $P = 15,829.32 - 1000 = 14,829.32$, $r = 0.0775$, and $t = 4$.

$$\begin{aligned} FV &= P(1+rt) \\ FV &= 14,829.32(1 + 0.0775 \cdot 4) \\ &= 19,426.41 \end{aligned}$$

There are 4 · 12 = 48 monthly payments, so the monthly payment for the car dealer's loan is 19,426.41/48 = 404.72.

The bank's loan is a 4-year simple interest amortized loan at 8.875% requiring a 10% down payment, or 0.10(15,829.32) = 1582.93. Thus $P = 15,829.32 - 1582.93 = 14,246.39$, $i = \frac{0.08875}{12}$, and $n = 48$ months.

$$\begin{aligned} pymt\frac{(1+i)^n - 1}{i} &= P(1+i)^n \\ pymt\frac{\left(1+\frac{0.08875}{12}\right)^{48} - 1}{\frac{0.08875}{12}} &= 14,246.39\left(1+\frac{0.08875}{12}\right)^{48} \\ pymt(57.3726851344) &= 20,291.4013533 \\ pymt &= \frac{20,291.4013533}{57.3726851344} \\ &= 353.68 \end{aligned}$$

The monthly payment for the bank loan is $353.68.

13b. Total interest for the car dealer's loan is $19,426.41 - 14,829.32 = 4597.09$, and for the bank loan is $48(353.68) - 14,246.39 = 2730.25$.

17a.

$$\begin{aligned} pymt\frac{(1+i)^n - 1}{i} &= P(1+i)^n \\ pymt\frac{\left(1+\frac{0.06}{12}\right)^{360} - 1}{\frac{0.06}{12}} &= 100,000\left(1+\frac{0.06}{12}\right)^{360} \\ pymt(1004.51504245) &= 602,257.521226 \\ pymt &= \frac{602,257.521226}{1004.51504245} \\ &= 599.55 \end{aligned}$$

Total interest $= 360(599.55) - 100,000 = 115,838.$

17b.

$$\begin{aligned}
pymt\frac{(1+i)^n-1}{i} &= P(1+i)^n \\
pymt\frac{\left(1+\frac{0.07}{12}\right)^{360}-1}{\frac{0.07}{12}} &= 100,000\left(1+\frac{0.07}{12}\right)^{360} \\
pymt(1219.97099412) &= 811,649.746568 \\
pymt &= \frac{811,649.746568}{1219.97099412} \\
&= 665.30
\end{aligned}$$

Total interest $= 360(665.30) - 100,000 = 139,508.$

17c.

$$\begin{aligned}
pymt\frac{(1+i)^n-1}{i} &= P(1+i)^n \\
pymt\frac{\left(1+\frac{0.08}{12}\right)^{360}-1}{\frac{0.08}{12}} &= 100,000\left(1+\frac{0.08}{12}\right)^{360} \\
pymt(1490.35945062) &= 1,093,572.96708 \\
pymt &= \frac{1,093,572.96708}{1490.35945062} \\
&= 733.76
\end{aligned}$$

Total interest $= 360(733.76) - 100,000 = 164,153.60.$

17d.

$$\begin{aligned}
pymt\frac{(1+i)^n-1}{i} &= P(1+i)^n \\
pymt\frac{\left(1+\frac{0.09}{12}\right)^{360}-1}{\frac{0.09}{12}} &= 100,000\left(1+\frac{0.09}{12}\right)^{360} \\
pymt(1830.74348307) &= 1,473,057.6123 \\
pymt &= \frac{1,473,057.6123}{1830.74348307} \\
&= 804.62
\end{aligned}$$

Total interest $= 360(804.62) - 100,000 = 189,663.20.$

17e.

$$\begin{aligned}
pymt\frac{(1+i)^n-1}{i} &= P(1+i)^n \\
pymt\frac{\left(1+\frac{0.10}{12}\right)^{360}-1}{\frac{0.10}{12}} &= 100,000\left(1+\frac{0.10}{12}\right)^{360} \\
pymt(2260.48792196) &= 1,983,739.93497 \\
pymt &= \frac{1,983,739.93497}{2260.48792196} \\
&= 877.57
\end{aligned}$$

Total interest $= 360(877.57) - 100,000 = 215,925.20.$

17f.

$$\begin{aligned}
pymt\frac{(1+i)^n-1}{i} &= P(1+i)^n \\
pymt\frac{\left(1+\frac{0.11}{12}\right)^{360}-1}{\frac{0.11}{12}} &= 100,000\left(1+\frac{0.11}{12}\right)^{360} \\
pymt(2804.51973988) &= 2,670,809.76156 \\
pymt &= \frac{2,670,809.76156}{2804.51973988} \\
&= 952.32
\end{aligned}$$

Total interest $= 360(952.32) - 100,000 = 242,835.20.$

21a.

$$\begin{aligned}
pymt\frac{(1+i)^n-1}{i} &= P(1+i)^n \\
pymt\frac{\left(1+\frac{0.10}{12}\right)^{360}-1}{\frac{0.10}{12}} &= 100,000\left(1+\frac{0.10}{12}\right)^{360} \\
pymt(2260.48792196) &= 1,983,739.93497 \\
pymt &= \frac{1,983,739.93497}{2260.48792196} \\
&= 877.57
\end{aligned}$$

Total interest $= 360(877.57) - 100,000 = 215,925.20.$

21b.

$$\begin{aligned}
pymt\frac{(1+i)^n-1}{i} &= P(1+i)^n \\
pymt\frac{\left(1+\frac{0.10}{26}\right)^{780}-1}{\frac{0.10}{26}} &= 100,000\left(1+\frac{0.10}{26}\right)^{780} \\
pymt(4932.27463919) &= 1,997,028.70738 \\
pymt &= \frac{1,997,028.70738}{4932.27463919} \\
&= 404.89
\end{aligned}$$

Total interest $= 780(404.89) - 100,000 = 215,814.20.$

25a.

$$\begin{aligned}
pymt\frac{(1+i)^n-1}{i} &= P(1+i)^n \\
pymt\frac{\left(1+\frac{0.0925}{12}\right)^{4}-1}{\frac{0.0925}{12}} &= 48,000\left(1+\frac{0.0925}{12}\right)^{4} \\
pymt &= 12,232.14
\end{aligned}$$

25b.

Payment No.	Principal Por.	Interest Por.	Tot. Payment	Balance
0				48,000.00
1	11,862.14	370.00	12,232.14	36,137.86
2	11,953.58	278.56	12,232.14	24,184.28
3	12,045.72	186.42	12,232.14	12,138.56
4	12,138.58	93.57	12,232.13	0

29. First find the monthly payment:

$$pymt\frac{(1+i)^n-1}{i} = P(1+i)^n$$

$$pymt\frac{\left(1+\frac{0.115}{12}\right)^{48}-1}{\frac{0.115}{12}} = 14,502.44\left(1+\frac{0.115}{12}\right)^{48}$$

$$pymt = 378.35$$

Now find the unpaid balance using the Unpaid Balance Formula. Here $n = 36 + 2 = 38$ (corresponding to monthly payments over 3 years plus 2 months).

$$P(1+i)^n - pymt\frac{(1+i)^n-1}{i} = 14,502.44\left(1+\frac{0.115}{12}\right)^{38} - 378.35\frac{\left(1+\frac{0.115}{12}\right)^{38}-1}{\frac{0.115}{12}}$$

$$\approx 14,502.44(1.43682062048) - 378.35(45.5812821371)$$

$$= 3591.73$$

33a. Monthly payment:

$$pymt\frac{(1+i)^n-1}{i} = P(1+i)^n$$

$$pymt\frac{\left(1+\frac{0.13375}{12}\right)^{360}-1}{\frac{0.13375}{12}} = 152,850\left(1+\frac{0.13375}{12}\right)^{360}$$

$$pymt = 1735.74$$

33b. The loan amount for the new loan will be the unpaid balance on the old loan which had been paid up to 3 years (thus $n = 36$):

$$P(1+i)^n - pymt\frac{(1+i)^n-1}{i} = 152,850\left(1+\frac{0.13375}{12}\right)^{36} - 1735.74\frac{\left(1+\frac{0.13375}{12}\right)^{36}-1}{\frac{0.13375}{12}}$$

$$\approx 152,850(1.49037884363) - 1735.74(43.9966065314)$$

$$= 151,437.74$$

33c.

$$pymt\frac{(1+i)^n-1}{i} = P(1+i)^n$$

$$pymt\frac{\left(1+\frac{0.08875}{12}\right)^{360}-1}{\frac{0.08875}{12}} = 151,437.74\left(1+\frac{0.08875}{12}\right)^{360}$$

$$pymt = 1204.91$$

33d. $360(1735.74) - 152,850 = 472,016.40$

33e. They have made 36 payments, and if they refinance they do not have to make $360 - 36 = 324$ more payments with a total interest of $324(1735.74) - 151,437.74 = 410,942.02$. If they do not refinance the total interest would be $360(1735.74) - 152,850 = 472,016.40$. Thus far (after 3 years) their total interest is $472,016.40 - 410,942.02 = 61,074.38$. Add that to the interest they will pay through the new loan, which is $360(1204.91) - 151,437.74 = 282,329.86$, and the total interest they pay if they get a new loan is $61,074.38 + 282,329.86 = 343,404.24$.

37a.

$$pymt\frac{(1+i)^n-1}{i}=P(1+i)^n$$

$$pymt\frac{\left(1+\frac{0.075}{12}\right)^{360}-1}{\frac{0.075}{12}}=100,000\left(1+\frac{0.075}{12}\right)^{360}$$

$$pymt=699.21$$

37b.

$$pymt\frac{(1+i)^n-1}{i}=P(1+i)^n$$

$$pymt\frac{\left(1+\frac{0.05375}{12}\right)^{360}-1}{\frac{0.05375}{12}}=100,000\left(1+\frac{0.05375}{12}\right)^{360}$$

$$pymt=559.97$$

37c. $12(699.21-559.97)=1670.88$

37d. Use the Unpaid Balance Formula:

$$\begin{aligned}P(1+i)^n-pymt\frac{(1+i)^n-1}{i}&=100,000\left(1+\frac{0.05375}{12}\right)^{12}-559.97\frac{\left(1+\frac{0.05375}{12}\right)^{12}-1}{\frac{0.05375}{12}}\\&\approx 100,000(1.05509412472)-559.97(12.3000836584)\\&=98,621.73\end{aligned}$$

37e. The loan's rate isn't allowed to rise more than 2% in any one adjustment, and it has a "ceiling" or upper bound of 11.875%. That means that in the first adjustment the rate cannot be more than 5.375% + 2% = 7.375%. Now, the current cost of funds of the 11th District Federal Home Loan Bank is 4.839%, and the adjustment adds 2.875% to that, equalling a rate of 4.839% + 2.875% = 7.714%. However, this exceeds 7.375%, so the correct rate is 7.375%.

Since one year has already elapsed, there are 29 years left to go on the loan, which means that $n=12\cdot 29=348$.

37f. Note that we must use $i=\frac{0.07375}{12}$, $n=348$, and $P=98,621.73$ (from the unpaid balance in Exercise 37d):

$$pymt\frac{(1+i)^n-1}{i}=P(1+i)^n$$

$$pymt\frac{\left(1+\frac{0.07375}{12}\right)^{348}-1}{\frac{0.07375}{12}}=98,621.73\left(1+\frac{0.07375}{12}\right)^{348}$$

$$pymt=687.65$$

37g. $24(699.21)-[12(687.65)+12(559.97)]=1809.60$

Section 10.5 Annual Percentage Rate on a Graphing Calculator

1. First we compute the monthly payment. Since he pays a 10% down payment,
$P = \$16,113.82 - 0.10(\$16,113.82) = \$16,113.82(1-0.10) = \$16,113.82(0.90) = \$14,502.44.$ Also $i = \frac{0.115}{12}$ and $n = 4 \cdot 12 = 48.$

$$pymt\frac{(1+i)^n - 1}{i} = P(1+i)^n$$

$$pymt\frac{\left(1+\frac{0.115}{12}\right)^{48} - 1}{\frac{0.115}{12}} = 14,502.44\left(1+\frac{0.115}{12}\right)^{48}$$

$$pymt = 378.35$$

The legal loan amount is *loan amount – points and fees* $= 14,502.44 - 814.14 = 13,688.30.$ We use the Simple Interest Amortized Loan Formula from above,

$$pymt\frac{(1+i)^n - 1}{i} = P(1+i)^n,$$

let $pymt = 378.35$, $n = 48$, $P = 13,688.30$, $i = x$, and graph each side of the equation simultaneously. That is, we graph $Y_1 = 378.35\frac{(1+x)^{48}-1}{x}$ and $Y_2 = 13,688.30(1+x)^{48}$ on the same set of axes:

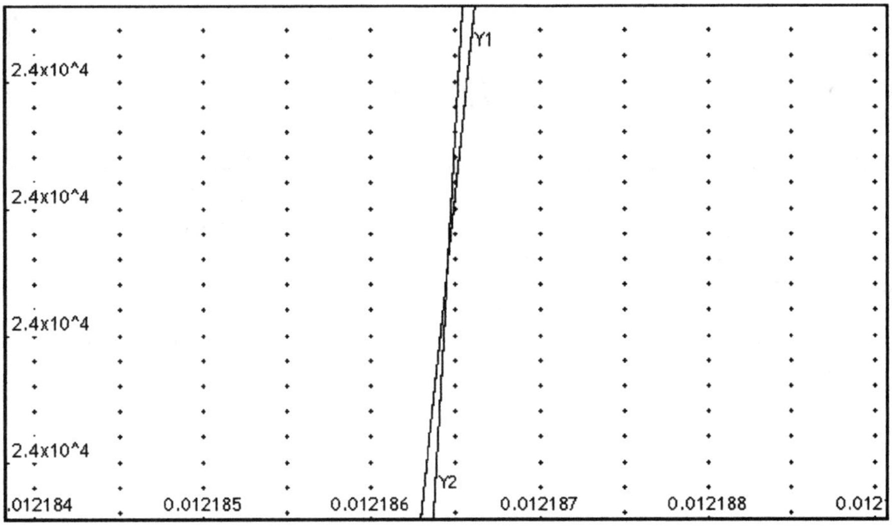

The intersection is at $i \approx 0.012186$; this is a monthly rate, and the corresponding annual rate is $12 \cdot 0.012186 \approx 0.1462 = 14.62\%$. This is the A.P.R.

5a. It is given that $P = 4600$, $r = 0.08$, and $t = 4$ years. The total interest charge is

$$\begin{aligned} I &= Prt \\ &= (4600)(0.08)(4) \\ &= 1472. \end{aligned}$$

The sum of principal and interest (the total cost of the loan) is $4600 + 1472 = 6072$. This is equally distributed over $4 \cdot 12 = 48$ monthly payments, so each monthly payment is $\frac{6072}{48}$ = \$126.50.

5b. We use the Simple Interest Amortized Loan Formula,

$$pymt\frac{(1+i)^n - 1}{i} = P(1+i)^n,$$

let $pymt = 126.50$, $n = 48$, $P = 4600$, $i = x$, and graph each side of the equation simultaneously. That is, we graph $Y_1 = 126.50\frac{(1+x)^{48}-1}{x}$ and $Y_2 = 4600(1+x)^{48}$ on the same set of axes:

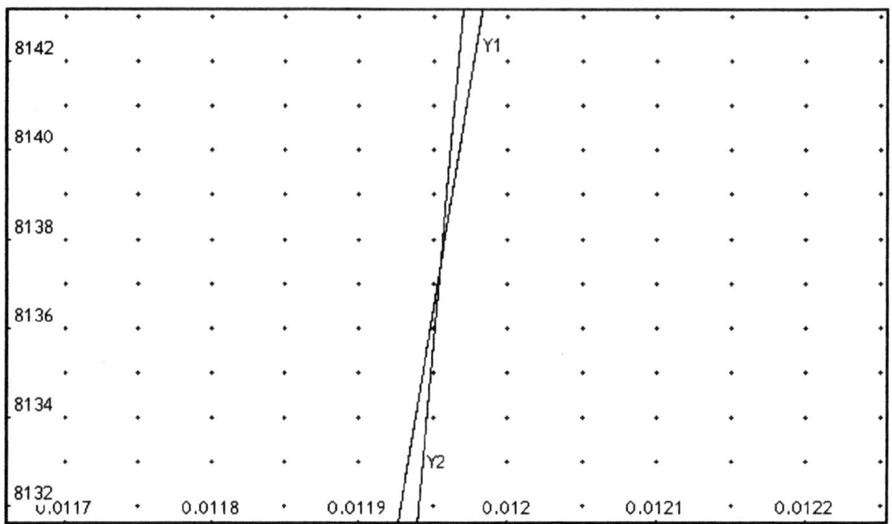

The intersection is at $i \approx 0.01195$; this is a monthly rate, and the corresponding annual rate is $12 \cdot 0.01195 \approx 0.143 = 14.3\%$. This is the A.P.R. Now, 14.3% – 14.25% (the claimed A.P.R.) = 0.05% < 0.125%, so the claimed A.P.R. is within the legal tolerance.

9. Since points and fees can vary among loaning institutions and thus affect the A.P.R. for each loan, this question cannot be unambiguously answered.

45a. Here we use a spreadsheet. The following shows the cells with their formulas, representing payments 1-12 and 349-360:

	Prin. Por.	**Int. Por.**	**Tot. Pymnt.**	**Balance**
0				170000
=A2+1	=D3-C3	=E2*0.10875/12	1602.91	=E2-B3
=A3+1	=D4-C4	=E3*0.10875/12	1602.91	=E3-B4
=A4+1	=D5-C5	=E4*0.10875/12	1602.91	=E4-B5
=A5+1	=D6-C6	=E5*0.10875/12	1602.91	=E5-B6
=A6+1	=D7-C7	=E6*0.10875/12	1602.91	=E6-B7
=A7+1	=D8-C8	=E7*0.10875/12	1602.91	=E7-B8
=A8+1	=D9-C9	=E8*0.10875/12	1602.91	=E8-B9
=A9+1	=D10-C10	=E9*0.10875/12	1602.91	=E9-B10
=A10+1	=D11-C11	=E10*0.10875/12	1602.91	=E10-B11
=A11+1	=D12-C12	=E11*0.10875/12	1602.91	=E11-B12
=A12+1	=D13-C13	=E12*0.10875/12	1602.91	=E12-B13
=A13+1	=D14-C14	=E13*0.10875/12	1602.91	=E13-B14

=A350+1	=D351-C351	=E350*0.10875/12	1602.91	=E350-B351
=A351+1	=D352-C352	=E351*0.10875/12	1602.91	=E351-B352
=A352+1	=D353-C353	=E352*0.10875/12	1602.91	=E352-B353
=A353+1	=D354-C354	=E353*0.10875/12	1602.91	=E353-B354
=A354+1	=D355-C355	=E354*0.10875/12	1602.91	=E354-B355
=A355+1	=D356-C356	=E355*0.10875/12	1602.91	=E355-B356
=A356+1	=D357-C357	=E356*0.10875/12	1602.91	=E356-B357
=A357+1	=D358-C358	=E357*0.10875/12	1602.91	=E357-B358
=A358+1	=D359-C359	=E358*0.10875/12	1602.91	=E358-B359
=A359+1	=D360-C360	=E359*0.10875/12	1602.91	=E359-B360
=A360+1	=D361-C361	=E360*0.10875/12	1602.91	=E360-B361
=A361+1	=E361	=E361*0.10875/12	=B362+C362	=E361-B362
Total Int.		**=SUM(C3:C362)**		

The following shows the values of each cell from those above:

	Prin. Por.	**Int. Por.**	**Tot. Pymnt.**	**Balance**
0				$170,000.00
1	$62.29	$1,540.63	$1,602.91	$169,937.72
2	$62.85	$1,540.06	$1,602.91	$169,874.87
3	$63.42	$1,539.49	$1,602.91	$169,811.45
4	$63.99	$1,538.92	$1,602.91	$169,747.45
5	$64.57	$1,538.34	$1,602.91	$169,682.88
6	$65.16	$1,537.75	$1,602.91	$169,617.72
7	$65.75	$1,537.16	$1,602.91	$169,551.97
8	$66.35	$1,536.56	$1,602.91	$169,485.63
9	$66.95	$1,535.96	$1,602.91	$169,418.68
10	$67.55	$1,535.36	$1,602.91	$169,351.13
11	$68.17	$1,534.74	$1,602.91	$169,282.96
12	$68.78	$1,534.13	$1,602.91	$169,214.18

349	$1,438.37	$164.54	$1,602.91	$16,717.74
350	$1,451.41	$151.50	$1,602.91	$15,266.34
351	$1,464.56	$138.35	$1,602.91	$13,801.78
352	$1,477.83	$125.08	$1,602.91	$12,323.95
353	$1,491.22	$111.69	$1,602.91	$10,832.72
354	$1,504.74	$98.17	$1,602.91	$9,327.99
355	$1,518.38	$84.53	$1,602.91	$7,809.61
356	$1,532.14	$70.77	$1,602.91	$6,277.48
357	$1,546.02	$56.89	$1,602.91	$4,731.46
358	$1,560.03	$42.88	$1,602.91	$3,171.42
359	$1,574.17	$28.74	$1,602.91	$1,597.26
360	$1,597.26	$14.48	$1,611.73	$0.00
Total Interest		**$407,056.42**		

45b. From the Interest Portion column of the spreadsheet, we add the data from the first 12 rows: 1540.63 + 1540.06 + ... + 1534.74 + 1534.13 = 18,449.10. This can also be done by inserting the sum command into the spreadsheet.

45c. From cell E350 of the spreadsheet: $18,156.11

45d.

348	$1,425.45	$177.46	$1,602.91	$18,156.11
349	$1,438.37	$164.54	$1,602.91	$16,717.74
350	$1,451.41	$151.50	$1,602.91	$15,266.34
351	$1,464.56	$138.35	$1,602.91	$13,801.78
352	$1,477.83	$125.08	$1,602.91	$12,323.95
353	$1,491.22	$111.69	$1,602.91	$10,832.72
354	$1,504.74	$98.17	$1,602.91	$9,327.99
355	$1,518.38	$84.53	$1,602.91	$7,809.61
356	$1,532.14	$70.77	$1,602.91	$6,277.48
357	$1,546.02	$56.89	$1,602.91	$4,731.46
358	$1,560.03	$42.88	$1,602.91	$3,171.42
359	$1,574.17	$28.74	$1,602.91	$1,597.26
360	$1,597.26	$14.48	$1,611.73	$0.00
Total Interest		**$407,056.42**		

45e. 164.54 + 151.50 + ... + 28.74 + 14.48 = 1087.62

49a. Here we use a spreadsheet. The following shows the cells with their formulas, representing payments 1-12 and 49-60:

Payment No.	Prin. Por.	Int. Por.	Tot. Pymnt.	Balance
0				10120
=A2+1	=D3-C3	=E2*0.07875/12	204.59	=E2-B3
=A3+1	=D4-C4	=E3*0.07875/12	204.59	=E3-B4
=A4+1	=D5-C5	=E4*0.07875/12	204.59	=E4-B5
=A5+1	=D6-C6	=E5*0.07875/12	204.59	=E5-B6
=A6+1	=D7-C7	=E6*0.07875/12	204.59	=E6-B7
=A7+1	=D8-C8	=E7*0.07875/12	204.59	=E7-B8
=A8+1	=D9-C9	=E8*0.07875/12	204.59	=E8-B9
=A9+1	=D10-C10	=E9*0.07875/12	204.59	=E9-B10
=A10+1	=D11-C11	=E10*0.07875/12	204.59	=E10-B11
=A11+1	=D12-C12	=E11*0.07875/12	204.59	=E11-B12
=A12+1	=D13-C13	=E12*0.07875/12	204.59	=E12-B13
=A13+1	=D14-C14	=E13*0.07875/12	204.59	=E13-B14
=A50+1	=D51-C51	=E50*0.07875/12	204.59	=E50-B51
=A51+1	=D52-C52	=E51*0.07875/12	204.59	=E51-B52
=A52+1	=D53-C53	=E52*0.07875/12	204.59	=E52-B53
=A53+1	=D54-C54	=E53*0.07875/12	204.59	=E53-B54
=A54+1	=D55-C55	=E54*0.07875/12	204.59	=E54-B55
=A55+1	=D56-C56	=E55*0.07875/12	204.59	=E55-B56
=A56+1	=D57-C57	=E56*0.07875/12	204.59	=E56-B57
=A57+1	=D58-C58	=E57*0.07875/12	204.59	=E57-B58
=A58+1	=D59-C59	=E58*0.07875/12	204.59	=E58-B59
=A59+1	=D60-C60	=E59*0.07875/12	204.59	=E59-B60
=A60+1	=D61-C61	=E60*0.07875/12	204.59	=E60-B61
=A61+1	=E61	=E61*0.07875/12	=B62+C62	=E61-B62
Total Int.			=SUM(C3:C62)	

The following shows the values of each cell from those above:

Payment No.	Prin. Por.	Int. Por.	Tot. Pymnt.	Balance
0				$10,120.00
1	$138.18	$66.41	$204.59	$9,981.82
2	$139.08	$65.51	$204.59	$9,842.74
3	$140.00	$64.59	$204.59	$9,702.74
4	$140.92	$63.67	$204.59	$9,561.83
5	$141.84	$62.75	$204.59	$9,419.98
6	$142.77	$61.82	$204.59	$9,277.21
7	$143.71	$60.88	$204.59	$9,133.51
8	$144.65	$59.94	$204.59	$8,988.85
9	$145.60	$58.99	$204.59	$8,843.25
10	$146.56	$58.03	$204.59	$8,696.70
11	$147.52	$57.07	$204.59	$8,549.18
12	$148.49	$56.10	$204.59	$8,400.69

49	$189.14	$15.45	$204.59	$2,164.49
50	$190.39	$14.20	$204.59	$1,974.11
51	$191.63	$12.96	$204.59	$1,782.47
52	$192.89	$11.70	$204.59	$1,589.58
53	$194.16	$10.43	$204.59	$1,395.42
54	$195.43	$9.16	$204.59	$1,199.99
55	$196.72	$7.87	$204.59	$1,003.27
56	$198.01	$6.58	$204.59	$805.27
57	$199.31	$5.28	$204.59	$605.96
58	$200.61	$3.98	$204.59	$405.35
59	$201.93	$2.66	$204.59	$203.42
60	$203.42	$1.33	$204.75	$0.00
Total Int.			$2,155.56	

49b. From the Interest Portion column of the spreadsheet, we add the data from the first 12 rows: 66.41 + 65.51 + ... + 57.07 + 56.10 = 735.7. This can also be done by inserting the sum command into the spreadsheet.

49c. This is found at the balance for payment 49: $2,164.49.

49d. (See 49a.)

49e. 14.20 + 12.96 + ... + 2.66 + 1.33 = 86.15

Chapter 10
Review Exercises

1.

$$\begin{aligned} I &= Prt \\ &= (8140)(0.0975)(11) \\ &= 8730.15 \end{aligned}$$

2.

$$\begin{aligned} FV &= P(1+i)^n \\ &= 8140\left(1+\frac{0.0975}{12}\right)^{11\cdot 12} \\ &= 8140(1.008125)^{132} \\ &\approx 23,687.72 \end{aligned}$$

Interest earned $= 23,687.72 - 8140 = 15,547.72.$

3. Note that 1 year and 2 months $= 1\frac{2}{12}$ years $= \frac{7}{6}$ years.

$$\begin{aligned} I &= Prt \\ &= (3550)(0.125)\left(\frac{7}{6}\right) \\ &\approx 4067.71 \end{aligned}$$

4.

$$\begin{aligned} FV &= P(1+i)^n \\ &= 3550\left(1+\frac{0.125}{12}\right)^{\frac{7}{6}\cdot 12} \\ &\approx 3550(1.01041666667)^{14} \\ &\approx 4104.26 \end{aligned}$$

5.

$$\begin{aligned} FV &= P(1+rt) \\ &= P + Prt \\ FV - P &= Prt \\ t &= \frac{FV - P}{Pr} \\ &= \frac{8000 - 7000}{(7000)(0.075)} \\ &\approx 1.9 \end{aligned}$$

6a. The amount of the promissory note is $0.15(205{,}500) = 30{,}825$. Using the formula $I = Prt$, the interest on this amount is $(30{,}825)(0.12)(8) = 29{,}592$. Since there are $12 \cdot 8 = 96$ monthly payment periods, the monthly payment is $\frac{29{,}592}{96} = 308.25$.

6b. At the time of sale Sanchez received $0.05(205{,}500) = 10{,}275$ down. The income from the promissory note is $30{,}825 + 29{,}592 = 60{,}417$. Thus the total amount Sanchez received from the down payment is $60{,}417 + 10{,}275 = 70{,}692$.

6c. Less the down payment of 20%, the amount to be paid for the house is 0.8(205,500) = 164,400. Adding this to the income from the down payment yields 164,400 + 70,692 = 235,092.

7. First we compute the annual yield for Extremely Trustworthy Savings:

$$\begin{aligned} FV(\text{compounded annually}) &= FV(\text{simple interest}) \\ P(1+i)^n &= P(1+rt) \\ (1+i)^n &= 1+rt \\ \left(1+\frac{0.0763}{1}\right)^1 &= 1+r\cdot 1 \\ 1.0763-1 &= r \\ r &= 0.0763 \end{aligned}$$

Now we compute the annual yield for Bank of the South:

$$\begin{aligned} FV(\text{compounded daily}) &= FV(\text{simple interest}) \\ P(1+i)^n &= P(1+rt) \\ (1+i)^n &= 1+rt \\ \left(1+\frac{0.0763}{365}\right)^{365} &= 1+r\cdot 1 \\ 1.07927770712-1 &= r \\ r &\approx 0.0793 \end{aligned}$$

Bank of the South is better for the consumer.

8a. We find the present value (principal P) which will generate a future value FV of \$250,000. The time involved is 65 − 25 = 40 years, and $n = 4\cdot 40 = 160$.

$$\begin{aligned} FV &= P(1+i)^n \\ P &= \frac{FV}{(1+i)^n} \\ &= \frac{250,000}{\left(1+\frac{0.10125}{4}\right)^{160}} \\ &\approx 4580.78 \end{aligned}$$

8b. When he retires the account balance is \$250,000, and every 30 days thereafter the interest continues to accrue, but he will withdraw just the interest at the end of every 30-day month, leaving \$250,000 in the account. Thus we need to find the interest which accrues every thirty days. That is, we find the future value of \$250,000 after 30 days, and subtract \$250,000 from it.

$$\begin{aligned} FV - 250,000 &= P(1+i)^n - 250,000 \\ &= 250,000\left(1+\frac{0.10125}{12}\right)^1 - 250,000 \\ &\approx 2109.38 \end{aligned}$$

9.

$$\begin{aligned} FV(\text{ordinary}) &= pymt\frac{(1+i)^n - 1}{i} \\ &= 200\frac{\left(1+\frac{0.06125}{12}\right)^{11\cdot 12} - 1}{\frac{0.06125}{12}} \\ &\approx 37,546.62 \end{aligned}$$

10.

$$\begin{aligned} FV &= P(1+i)^n \\ P &= \frac{FV}{(1+i)^n} \\ &= \frac{37,546.62}{\left(1+\frac{0.06125}{12}\right)^{132}} \\ &\approx 19,173.84 \end{aligned}$$

11.

$$\begin{aligned} FV(\text{due}) &= pymt\frac{(1+i)^n - 1}{i}(1+i) \\ &= 200\frac{\left(1+\frac{0.06125}{12}\right)^{132} - 1}{\frac{0.06125}{12}}\left(1+\frac{0.06125}{12}\right) \\ &\approx 37,738.26 \end{aligned}$$

12a.

$$\begin{aligned} 1300 &= \frac{0.0825(FV)}{12} \\ FV &= 1300 \cdot \frac{12}{0.0825} \\ &\approx 189,090.91 \end{aligned}$$

12b.

$$\begin{aligned} FV(\text{ordinary}) &= pymt\frac{(1+i)^n - 1}{i} \\ pymt &= \frac{iFV}{(1+i)^n - 1} \\ &= \frac{\frac{0.0975}{12} \cdot 189,090.91}{\left(1+\frac{0.0975}{12}\right)^{360} - 1} \\ &\approx 88.22 \end{aligned}$$

13a.

$$\begin{aligned} pymt\frac{(1+i)^n - 1}{i} &= P(1+i)^n \\ pymt &= \frac{iP(1+i)^n}{(1+i)^n - 1} \\ &= \frac{\left(\frac{0.10875}{12}\right)(12,138.58)\left(1+\frac{0.10875}{12}\right)^{60}}{\left(1+\frac{0.10875}{12}\right)^{60} - 1} \\ &\approx 263.17 \end{aligned}$$

13b. There are 60 payments of $263.17 each for a total of 15,790.20. Of this, 12,138.58 is principal, leaving the total interest of 3651.62.

13c.

Payment No.	Principal Por.	Interest Por.	Tot. Payment	Balance
0				12,138.58
1	153.16	110.01	263.17	11,985.42
2	154.55	108.62	263.17	11,830.87

13d. Loan amount – points and fees = 12,136.58 – 633.87 = 11,502.71.

$$pymt\frac{(1+i)^n-1}{i} = P(1+i)^n$$
$$263.17\frac{(1+i)^{60}-1}{i} = 11,502.71(1+i)^{60}$$

Let $Y_1 = 263.17\frac{(1+x)^{60}-1}{x}$, $Y_2 = 11,502.71(1+x)^{60}$. Graph these simultaneously. Their intersection is at $i = 0.0110379362901$; this is a monthly rate. The annual rate is $12 \cdot 0.0110379362901 \approx 0.1325 = 13.25\%$. (This is the A.P.R.) However, 13.25% – 10% = 3.25% > 0.125%, so the lender's statement of A.P.R. does not verify.

13e. Note that n below equals 2(12) + 6, or 30.

$$\begin{aligned}\text{unpaid balance} &= P(1+i)^n - pymt\frac{(1+i)^n-1}{i}\\ &= 12,136.58\left(1+\frac{0.10875}{12}\right)^{30} - 263.17\frac{\left(1+\frac{0.10875}{12}\right)^{30}-1}{\frac{0.10875}{12}}\\ &\approx 6885.51\end{aligned}$$

14a.

$$\begin{aligned}I &= Prt\\ &= (6000)(0.099)(4)\\ &= 2376\end{aligned}$$

The total of principal and interest is $P + I = 6000 + 2376 = 8376$. This is equally distributed among 48 monthly payments, so each monthly payment is $\frac{8376}{48} = 174.50$.

14b.

$$pymt\frac{(1+i)^n-1}{i} = P(1+i)^n$$
$$174.50\frac{(1+i)^{48}-1}{i} = 6000(1+i)^{48}$$

Let $Y_1 = 174.50\frac{(1+x)^{48}-1}{x}$, $Y_2 = 6000(1+x)^{48}$. Graph these simultaneously. Their intersection is at $i = 0.0145335705585$; this is a monthly rate. The annual rate is $12 \cdot 0.0145335705585 \approx 0.1744 = 17.44\%$. (This is the A.P.R.) However, 17.44% – 10% = 7.44% > 0.125%, so the lender's statement of A.P.R. does not verify.

33. $\frac{\frac{9-12}{5}+7}{2} = 3.2$

34. $\frac{\frac{4-11}{6}+8}{7} \approx 0.97619047619$

35. $\frac{\frac{7+9}{5}+\frac{8-14}{3}}{3} = 0.4$

36. $\frac{\frac{4-16}{5}+\frac{7-22}{2}}{5} = -1.98$

37. $x \approx -0.142857142857$ or $x \approx -1.83333333333$

38. $x \approx 0.528902144254$ or $x \approx -2.6867968811$

41a. $\frac{\frac{5+7.1}{3}+\frac{2-7.1}{5}}{7} \approx 0.430476190476$

41b. $\frac{\frac{5+7.2}{3}+\frac{2-7.2}{5}}{7} \approx 0.432380952381$

41c. $\frac{\frac{5+9.3}{3}+\frac{2-9.3}{5}}{7} \approx 0.472380952381$

42a. $\frac{3+\frac{5+\frac{6-8.3}{2}}{3}}{9} \approx 0.475925925926$

42b. $\frac{3+\frac{5-\frac{6-8.3}{2}}{3}}{9} \approx 0.561111111111$

42c. $\frac{3-\frac{5+\frac{6-8.3}{2}}{3}}{9} \approx 0.190740740741$